Why Teach Physics?

Why Teach Physics?

BASED ON DISCUSSIONS AT THE
INTERNATIONAL CONFERENCE ON
PHYSICS IN GENERAL EDUCATION

*Palácio da Cultura
Rio de Janeiro · Brazil
July 1–6 · 1963*

Edited by
SANBORN C. BROWN
NORMAN CLARKE
JAYME TIOMNO

*Published for
The International Union of Pure and Applied Physics
by*
THE M.I.T. PRESS
*The Massachusetts Institute of Technology
Cambridge, Massachusetts*

MIT Press

0262020084

BROWN
WHY TEACH PHYSICS

Preface

This volume contains the editors' survey of the proceedings of a Conference on Physics in General Education held in Rio de Janeiro in July, 1963. This was the second conference on education to be organized under the auspices of the International Union of Pure and Applied Physics (IUPAP), in association with a number of other national and international organizations. The first conference was held in UNESCO HOUSE in Paris in the summer of 1960.* The need for the first conference had become apparent from informal discussions among physicists of several countries, and the conference was organized by an ad hoc international committee; a great deal of interest was aroused throughout the world, and attendance was restricted to a limited number of officially appointed delegates from the major organizations of physics in each country. No fewer than twenty-eight countries and four international organizations were officially represented, and among them were all those countries in which physics is highly developed as a part of education and as an essential foundation of technology.

At the end of the Conference a number of resolutions were unanimously adopted. One of these was that the International Union of Pure and Applied Physics should be asked to set up a permanent committee which would be responsible for keeping under review and for actively stimulating the development of physics education internationally. Later in 1960 this recommendation was adopted by the International Union at its general assembly in Ottawa, and an International Commission on Physics Education was appointed with the following membership:

Professor Sanborn C. Brown (U.S.A.), President
Mr. Norman Clarke (U.K.), Secretary
Professor Pierre Fleury (France)
Professor V. S. Fursov (succeeded by Professor A. S. Akhmatov) (U.S.S.R.)
Professor A. M. J. F. Michels (The Netherlands)
Professor D. Sette (Italy)
Professor Jayme Tiomno (Brazil)

---

*International Education in Physics, Sanborn C. Brown and Norman Clarke, Eds., The Technology Press and John Wiley & Sons, Inc., New York, 1960.

and with two corresponding members, Dr. M. A. El-Sherbini (Egypt) and Professor M. Valouch (Czechoslovakia). Since then the Commission has initiated a number of relevant activities including a regular, formal arrangement whereby it advises UNESCO on matters concerning physics education.

A second recommendation of the Paris Conference, also accepted by the general assembly of the International Union, was that further conferences should be held on specific fields of physics education. The Paris Conference, perhaps inevitably, had covered a very wide field, although it was by no means concerned with generalities, and a great deal of detailed, specific information had been submitted in previously prepared papers discussed at the conference. Nevertheless, it was felt that there was a great need to discuss in detail some of the points that had emerged in Paris, and also to relate some of the topics to the particular problems of different areas of the world. The International Commission, therefore, decided that the second conference, which is here summarized, should be concerned with problems that arise in the teaching of physics as part of the general education of all children. The first of the resolutions passed in Paris has, indeed, begun in the following words:

> "In our view, physics is an essential part of the intellectual life of man at the present day," and had gone on to say "Studying physics and the physicist's methods of acquiring and evaluating knowledge should therefore be regarded as a necessary part of the education of all children."

In those countries where science is most highly developed, this has been accepted for many years as a desirable aim, and attempts have been made with varying degrees of success to implement it. In such countries difficulties of technique have been met and, perhaps above all, the difficulty of deciding the appropriate content of a course of study of physics intended to be part of the liberal education of children who are not going to be science specialists.

There are, however, many countries of the world with educational systems that do not regard science as more than a useful activity for those who are perhaps not of the highest intellectual ability, or for those who at an early age have decided that they will take up occupations for which a knowledge of science is essential. In the main, these are countries which are not highly developed either scientifically or economically, and the Commission recognized that a conference on this subject, if directed to the needs of such countries, would have to devote a substantial part of its time to the presentation of the basic arguments for the inclusion of science and of physics, in particular, in the education of all children.

For a number of reasons and after consultation with UNESCO, the Commission decided that the Conference should be held in Latin America and was glad to accept an invitation to hold it in Rio de Janeiro; this proved to be a good choice both because of relatively easy accessibility from other Latin American countries and because of the attractiveness of the city itself. The active support of the Organization of the American States, the Brazilian Ministry of Education and Culture, the

Brazilian National Research Council, the Latin American Center for Physics, and the Brazilian Center for Physics Research was readily proffered. This support was far more than formal, and the international character of the Latin American Center for Physics Research proved to be of invaluable assistance. A Committee under the chairmanship of Professor Jayme Tiomno was set up to deal with the local organization and made excellent arrangements.

It was obvious that a conference in South America would face some special organizational difficulties and would have to be arranged quite differently from that held in Paris. The Paris Conference had been organized from London, with very close, extensive, and experienced help from the French Physical Society and the Secretariat of UNESCO. In Europe and in the United States, organizers of large conferences have within easy call almost unlimited material and scientific resources and a virtually unlimited amount of experienced assistance and advice. If, for example, they wish to arrange an exhibition, there is no shortage of research laboratories and manufacturers who can provide almost anything that is required. Moreover, since Europe, the United States, and Russia are the main centers of scientific activity, it is evident that meetings held there will attract substantial numbers of scientists of distinction as participants. The less developed countries of the world are obviously unable to offer such clear advantages, but the International Commission believes that much of its work should be arranged and located so as to be of special interest to the countries that, scientifically, are less developed. Partly to meet the problem of ease of attendance, the Commission was very glad to cooperate with the Organization of American States, which was itself planning a conference on physics to be held in Latin America. By a very happy arrangement, the OAS conference was held in Rio de Janeiro in the week immediately preceding the Conference on Physics Education, and it was thus possible for a large number of physicists from South America to participate in both conferences at the cost of only one journey.

For the reasons which have been indicated, the main differences between the Paris and Rio de Janeiro Conferences were that in the second one it was not possible to have as adequate a participation from countries outside Latin America; it was not possible to have the same range of prepared papers; import formalities and restrictions made it impossible to have a truly international exhibition of equipment, and no exhibition of books could be arranged.

It was decided to present the proceedings in the same broad way as those of the Paris Conference, namely, as a volume that represented the impressions of the editors and was not merely a formal record of papers presented and the discussion upon them. At the end of the conference, the editors had the unusual experience of being, with their clerical assistants, the only occupants of a 50-bedroom hotel which from a height of 2000 feet overlooks Rio de Janeiro; such solitude for a task of this kind might well be difficult to achieve in more developed parts of the world!

The reader may consider it surprising that the record of the Paris Conference, on a much more diffuse theme, contains a great deal of detailed information of practical value to the teacher of physics, whereas the record of this conference emphasizes, in the main, general problems of educational policy. The reason is that the discussion at this conference was along lines dictated by the problems and attitudes of South America, where it is still necessary to argue in policy terms for the inclusion of physics in general education.

The editors have condensed and in some cases rewritten very considerably the material presented, and on only one or two matters has it been thought necessary to record any summary of discussion. In the addresses and papers there is a certain amount of repetition of material presented from slightly different viewpoints, and the editors were advised that it would be useful for the purposes of Latin America that these different points of view should remain in the published record.

We are grateful to the Ealing Corporation of Cambridge, Massachusetts, for financial aid in producing and distributing these proceedings.

Finally, the editors wish to record personal appreciation of the help given to them by Mrs. Lois Brown in editing some of the material and by Mrs. Margaret Buescu and Mr. Gerardo Eustace, who were invaluable in translating some material and in typing the manuscript.

Hotel Corcovado	Sanborn C. Brown
"Paineiras"	Norman Clarke
Rio de Janeiro	Jayme Tiomno
July 16, 1963	

Contents

Organization

International Conference on
Physics in General Education

— 1963 —

Organized by the Commission on Physics Education of the
International Union of Pure and Applied Physics

under the auspices of

International Union of Pure and Applied Physics
UNESCO
Organization of American States
Ministério da Educacão e Cultura
Conselho Nacional de Pesquisas
Centro Latino Americano de Física
Centro Brasileiro de Pesquisas Físicas

Chairmen:

Professor Sanborn C. Brown   Professor Jayme Tiomno

Commission on Physics Education of IUPAP:

Professor Sanborn C. Brown (U.S.A.) Chairman
Mr. Norman Clarke (United Kingdom) Secretary
Professor Pierre Fleury (France)
Professor A. M. J. F. Michels (The Netherlands)
Professor D. Sette (Italy)
Professor Jayme Tiomno (Brazil)
Professor A. S. Akhmatov (U.S.S.R.)
Dr. M. A. El-Sherbini (Egypt)
Professor M. Valouch (Czechoslovakia)

Local Commission:

Professor Jayme Tiomno
Professor Gabriel Fialho
Dr. Isaias Raw
Professor Oscar Sala
Professor Paulus Aulus Pompéia
Professor Francisco Magalhães Gomes

List of Participants

ARGENTINA

M. E. Báncora, Universidad Nacional del Litoral, Libertador General San Martín 8250, Buenos Aires

F. Cernuschi, Departamento de Física, Facultad de Ingeniería, Paseo Colon 850, Buenos Aires

C. A. Frumento, Facultad de Agronomía, Buenos Aires

E. E. Galloni, Facultad de Ingeniería, Paseo Colón 850, Buenos Aires

J. J. Giambiagi, Universidad de Buenos Aires, Buenos Aires

I. Kimel, Facultad de Ingeniería, Buenos Aires (presently at the Centro Brasileiro de Pesquisas Físicas, Av. Wenceslau Braz 71, Rio de Janeiro, GB, Brazil)

H. C. Panepucci, Facultad de Ciencias Exactas y Naturales (presently at the Centro Brasileiro de Pesquisas Físicas, Av. Wenceslau Braz 71, Rio de Janeiro, GB , Brazil)

J. G. Roederer, Universidad de Buenos Aires, Perú 272, Buenos Aires

BOLIVIA

I. Escobar V. Laboratorio de Física Cósmica, Universidad Mayor de San Andrés, La Paz

R. Vidaurre, Laboratorio de Física Cósmica, Universidad Mayor de San Andrés, La Paz

BRAZIL

R. M. X. de Araújo, Centro Brasileiro de Pesquisas Físicas, Av. Wenceslau Braz 71, Rio de Janeiro, GB

B. Alvarenga, Escola de Engenharia da Universidade de Minas Gerais, Rua Bolívar 45, Belo Horizonte, MG

R. M. Argollo, Centro Brasileiro de Pesquisas Físicas, Av. Wenceslau Braz 71, Rio de Janeiro, GB

A. Armando, Escola de Engenharia, Fortaleza, Ceará

G. B. de Souza Avila, Escola Politécnica da Universidade da Bahia, Rua Aristides Nevis 2, Salvador, Ba.

P. M. Barbosa, Colégio Antônio Vieira, Formiga, MG

O. Rocha Baiocchi, Instituto de Física da Universidade de Rio Grande do Sul, Pôrto Alegre, RGS

N. Bernardes, Faculdade de Filosofia, Ciências e Letras da Universidade de São Paulo, C. P. 8105, São Paulo

(Brazil — cont'd)

J. W. Bautista Vidal, Escola Politécnica, Bahia—Federação, Salvador, Ba.

F. Bunchaft, Faculdade Nacional de Filosofia, Av. Antônio Carlos 40, Rio de Janeiro, GB

A. Brotas, Instituto de Física e Matemática, Rua Progresso 465, Recife, Pe.

H. Carvalho, Centro Brasileiro de Pesquisas Físicas, Av. Wenceslau Braz 71, Rio de Janeiro, GB

N. Castro Faria, Centro Brasileiro de Pesquisas Físicas, Av. Wenceslau Braz 71, Rio de Janeiro, GB

R. Caniato, Universidade Católica, Campinas, S.P.

C. Borghi, Instituto de Física e Matemática, Rua do Progresso 465, Recife, Pe.

S. Cuyabano de Barros, Centro Brasileiro de Pesquisas Físicas, Av. Wenceslau Braz 71, Rio de Janeiro, GB

R. Carvalho, Instituto de Física da Universidade do Rio Grande do Sul, Pôrto Alegre, RGS

T. L. Cullen, Pontifícia Universidade Católica, Rua Marques de São Vicente 209, Rio de Janeiro, GB

R. Bastos da Costa, Centro Latino-Americano de Física, Av. Wenceslau Braz 71, Rio de Janeiro, GB

M. Côrtes Pires, Colégio Estadual, Caratinga, Minas Gerais

C. A. Dias, Centro Brasileiro de Pesquisas Físicas, Av. Wenceslau Braz 71, Rio de Janeiro, GB

C. Zaki Dib, Faculdade de Filosofia, Ciências e Letras da Universidade de São Paulo, C. P. 8105, S. P.

C. M. Ferreira Chaves, Faculdade Nacional de Filosofia, Rio de Janeiro, GB

G. Fialho, Centro Brasileiro de Pesquisas Físicas, Av. Wenceslau Braz 71, Rio de Janeiro, GB

E. Ferreira, Centrol Brasileiro de Pesquisas Físicas, Av. Wenceslau Braz 71, Rio de Janeiro, GB

C. Figueiredo Vieira, Centro Brasileiro de Pesquisas Físicas, Av. Wenceslau Braz 71, Rio de Janeiro, GB

E. Frota Pessoa, Centro Brasileiro de Pesquisas Físicas e Faculdade Nacional de Filosofia, Rio de Janeiro, GB

A. Gerbasi da Silva, Centro Brasileiro de Pesquisas Físicas, Av. Wenceslau Braz 71, Rio de Janeiro, GB

B. Coelho Pontes, Instituto de Física da Universidade do Estado da Guanabara, Rio de Janeiro, GB

L. Dias de Moura, Reitor da Pontifícia Universidade Católica, R. Marques da São Vicente 209, Rio de Janeiro, GB

D. Goldman, Departamento de Física da Universidade de São Paulo, São Paulo

P. Gomes de Paula Leite, Escola Nacional de Geologia e Escola Nacional de Engenharia, Rio de Janeiro, GB

S. W. Mac Dowell, Centro Brasileiro de Pesquisas Físicas, Av. Wenceslau Braz 71, Rio de Janeiro, GB

F. de Assis Magalhães Gomes, Faculdade de Filosofia da Universidade de Minas Gerais, Belo Horizonte, MG

(Brazil — cont'd)

A. Marques de Oliveira, Centro Brasileiro de Pesquisas Físicas, Av. Wenceslau Braz 71, Rio de Janeiro, GB

M. Q. Moreno, Faculdade de Filosofia da Universidade de Minas Gerais, Rua Carangola 288, Belo Horizonte, MG

W. Medeiros Duarte, Colégio Pedro II, Campo de São Cristovão, Rio de Janeiro, GB

E. Mello de Carvalho, Colégio Batista Mineiro, R. Buarque de Macedo 17, Belo Horizonte, MG

C. O. Lopes de Mendonca, Colégio Estadual de João Pessoa, João Pessoa, Paraiba

D. Mirza Abraham, Faculdade de Filosofia, Ciências e Letras da Universidade do Estado do Rio de Janeiro, Niterói, Estado do Rio

L. Môura, Centro Brasileiro de Pesquisas Físicas, Av. Wenceslau Braz 71, Rio de Janeiro, GB

H. M. Nussenzveig, Centro Brasileiro de Pesquisas Físicas, A. Wenceslau Braz 71, Rio de Janeiro, GB

M. C. Levi Nussenzveig, Centro Brasileiro de Pesquisas Físicas, Av. Wenceslau Braz 71, Rio de Janeiro, GB

E. E. Nelson, Colégio Adventista Brasileiro, C. P. 7258, São Paulo, S. P.

R. A. Normando, Instituto de Física da Universidade de Ceará, Fortaleza, Ceará

E. Nunes Pereira, Colégio Estadual Visconde Cairu, Rua Soares, Rio de Janeiro, GB

E. de Vasconcellos Paes, Faculdade de Filosofia da Universidade de Minas Gerais, Belo Horizonte, MG

A. Passos Guimarães Filho, Centro Brasileiro de Pesquisas Físicas, Av. Wenceslau Braz 71, Rio de Janeiro, GB

P. A. Pompeia, Instituto Tecnológico da Aeronáutica, São José dos Campos, São Paulo

J. H. Ponte, Escola de Engenharia, Fortaleza, Ceará

J. J. de Salles Pupo, Instituto de Física da Pontifícia Universidade Católica, Rua Marques de São Vicente 169, Rio de Janeiro, GB

I. Raw, Instituto Brasileiro de Educação, Ciência e Cultura, São Paulo

O. R. Ritter, Colégio Adventista Brasileiro, C. P. 7258, São Paulo

M. Rômulo, Instituto de Física e Matemática, Recife, Pe.

F. X. Roser, Pontifícia Universidade Católica, Rua Marques de São Vicente 169, Rio de Janeiro, GB

B. Silva Pereira, Faculdade de Filosofia, Rua Carangola 288, Belo Horizonte, MG

O. Sala, Laboratório Van de Graaff, Universidade de São Paulo, São Paulo

R. Sampel, Colégio Estadual e Escola Normal de Ibitinga, Ibitinga, S. P.

O. Sanches, Faculdade de Filosofia, Ciência e Letras, Rio Claro, S. P.

E. F. Soares. Instituto de Física da Universidade de Ceará, Fortaleza, Ceará

H. G. de Souza, Faculdade de Filosofia, Ciência e Letras, Rio Claro, S. P. (presently at O.A.S. Pan American Union, Washington 6, D. C. U.S.A.)

C. Schmitz, Instituto de Física da Universidade de Rio Grande do Sul, Pôrto Alegre, RGS

(Brazil — cont'd)

L. E. da Silva Machado, Colégio Franco-Brasileiro e Faculdade Nacional de Filosofia, Rio de Janeiro, GB

E. Silva, Departamento de Física da Universidade de São Paulo, São Paulo

P. Srivastava, Centro Brasileiro de Pesquisas, Av. Wenceslau Braz 71, Rio de Janeiro, GB

R. Tamburini Jr., Colégio de Alfenas, Belo Horizonte, MG

J. Tiomno, Centro Brasileiro de Pesquisas Físicas, Av. Wenceslau Braz 71, Rio de Janeiro, GB

A. Vienken, Colégio Arnaldo, Belo Horizonte, MG

J. R. Vasconcellos, Colégio O Precursor, Av. Olegário Maciel 1627, Belo Horizonte, MG

F. Medeiros Vieira, Núcleo de Física da Universidade do Pará, Belém, Pará

M. Schenberg, Universidade de São Paulo, São Paulo

A. da Silva Ximenes, Centro Brasileiro de Pesquisas Físicas, Av. Wenceslau Braz 71, Rio da Janeiro, GB

J. B. Martins, Instituto Militar de Engenharia, R. Conde de Bomfim 1136, Rio de Janeiro, GB

V. Ceotto, Centro de Ensino Médio, Quadra 12 — H. P. Sul c/232, Brasília

J. S. Helman, Centro Brasileiro de Pesquisas Físicas, Rio de Janeiro, GB

W. C. de Moraes Bastos, Colégio Estadual Paulo de Frontin, R. Barão de Ubá 399, Rio de Janeiro, GB

D. Crossetti, Instituto de Física da Universidade de Santa Maria, Rio Grande do Sul

C. L. Garcia, Colégio Nilo Peçanha, Rua Benjamim Constant 18, Niteroi, Estado do Rio

P. M. Guimarães Ferreira, Colégio Santo Inácio, Rua São Clemente 226. Rio de Janeiro. GB

R. F. Marchesini, Colégio Estadual Barão de Rio Branco, Rua Teodoro da Silva 620, Rio de Janeiro, GB

G. da Rosa Corrêa, Colégio Estadual Prefeito Mendes de Moraes, Rua Visconde de Santa Isabel 79, Rio de Janeiro, GB

B. Gross, Pontifícia Universidade Católica, Rio de Janeiro, GB

CANADA

P. Lorrain, Université de Montréal, Montréal

CHILE

M. Robert, Universidad de Chile, Santiago

H. Muñoz, Universidad de Chile, Santiago (presently at the following address: C. P. 2921, São Paulo, Brazil)

D. Moreno, Instituto de Física, Universidad de Chile, Casilla 2777, Santiago

N. Joel, Instituto de Física y Matemáticas, Universidad de Chile, Santiago (presently at the Instituto Brasileiro de Educação, Ciência e Cultura C. P. 2921, São Paulo, Brazil)

(Chile — cont'd)

G. Melcher, Instituto de Física y Matemáticas, Casilla 2777, Santiago

G. Alvial, Centro de Radiación Cósmica, Universidad de Chile, Casilla 1314, Santiago

R. F. Hernández Paves, Escuela de Ingenieros Industriales, Universidad Técnica del Estado, Casilla 10233, Santiago (presently at the Centro Brasileiro de Pesquisas Físicas, Av. Wenceslau Braz 71, Rio de Janeiro, Brazil)

J. Westphal, Instituto Pedagógico de la Universidad de Chile (presently at the following address: C. P. 2921, São Paulo, Brazil)

COLOMBIA

G. Jiménez Escobar, Universidad Nacional de Colombia, Bogotá (presently at the Centro Brasileiro de Pesquisas Físicas, Av. Wenceslau Braz 71, Rio, Brazil)

J. Herkrath, Universidad Nacional de Colombia, Departamento de Física, Ap. N. 1537, Bogotá

COSTA RICA

F. González, Universidad de Costa Rica, San José

N. N. Clarke, Universidad de Costa Rica, San José (presently at the Centro Brasileiro de Pesquisas Físicas, Av. Wenceslau Braz 71, Rio de Janeiro, Brazil)

CZECHOSLOVAKIA

M. Valouch, Université Charles Praha, Ke Karlovo 5, Praha

DOMINICAN REPUBLIC

J. O. Reyes, Universidad Autónoma de Santo Domingo, Ciudad Universitaria, Santo Domingo

ECUADOR

A. A. Freire, Universidad Central, Casilla 2110, Quito

G. F. Sosa Borja, Universidad Central, Casilla 2110, Quito (presently at the Centro Brasileiro de Pesquisas Físicas, Av. Wenceslau Braz 71, Rio de Janeiro, GB, Brazil)

FRANCE

P. Fleury, Union Internationale de Physique, Institut d'Optique, 3 Boulevard Pasteur, Paris 15°

A. Payan, Ministère de l'Education Nationale, Paris

G. A. Boutry, Conservatoire National des Arts et Métiers, Paris

R. Annequin, Lycée Chaptal, 45 Boulevard des Batignolles, Paris

GUATEMALA

M. A. Canga Arguelles, Universidad de San Carlos de Guatemala, Ciudad Universitaria, Guatemala

HONDURAS

R. Domínguez Agurcia, Centro de Estudios Generales, Universidad Nacional Autónoma de Honduras, Tegucigalpa, D. C.

ITALY

D. Sette, University of Rome, Piazzale Scienze 7, Rome

JAPAN

S. Kittaka, Tokyo College of Science, Kagutaraka, Shinjuku-Ku, Tokyo
A. Harasima, International Christian University, Oosawa, Mitaka, Tokyo

MOROCCO

H. Arseliès, Faculté des Sciences, Rabat

MEXICO

F. Sánchez Sinencio, Instituto Politécnico Nacional, Mexico (presently at Centro Brasileiro de Pesquisas Físicas, Av. Wenceslau Braz 71, Rio de Janeiro, GB, Brazil)
V. Flores Maldonado, Escuela Superior de Física y Matemáticas del I.P.N., Edificio 6, Ciudad de Zacatenco, Mexico 14, D.F.
F. Medina Nicolau, Instituto de Física, Universidad Nacional Autónoma de México, Mexico.
J. M. Lozano, Universidad Nacional Autónoma de México y Comisión Nacional de Energía Nuclear, Mexico.

THE NETHERLANDS

C. A. M. Michels-Veraart, Middenweg 92, Amsterdam
A. M. J. F. Michels, University of Amsterdam, Amsterdam

NICARAGUA

R. Argeñal, Departamento de Física y Matemáticas, Universidad Nacional de Nicaragua, Leon

PANAMA

B. Lombardo, Universidad de Panamá, Panama

PERU

A. Bueno, Instituto de Física, Universidad de San Marcos, Lima
A. Vidal, Universidad Nacional de Trujillo, Trujillo (presently at the Centro Brasileiro de Pesquisas Físicas, Av. Wenceslau Braz 71, Rio de Janeiro, Brazil)
E. López Carranza, Universidad Nacional de Trujillo, Trujillo
P. León Chincha, Departamento de Física, Universidad Nacional Mayor de San Marcos, Av. Nicolás de Piérola, Lima
V. H. Saito, Universidad Nacional Mayor de San Marcos, Lima

PARAGUAY

J. C. Lebrón, Instituto de Ciencias, Casilla 1141, Asunción

SOUTH AFRICA

J. W. Brommert, Department of Physics, University of the Witwatersrand, Johannesburg

SPAIN

L. Bru, Universidad de Madrid, Ciudad Universitaria, Madrid

SWEDEN

E. P. G. Ingelstam, Royal Institute, of Technology, Stockholm, 70
K. G. Friskopp, Johannesbergig 8, Orebro

UNITED KINGDOM

J. J. Lewis, Malvern College, Malvern, Worcestershire
N. Clarke, Institute of Physics and Physical Society, 47 Belgrave Square, London, S. W. 1
E. M. Rogers, 13A Princes Crescent, Hove 3, Sussex

UNITED STATES

S. C. Brown, Massachusetts Institute of Technology, Cambridge, Mass.
R. P. Fevnman, Department of Physics, California Institute of Technology, Pasadena, Calif.
G. Holton, Harvard University, Cambridge 38, Mass.
W. C. Kelly, American Institute of Physics, 335 East 45th Street, New York 17, N.Y.
J. R. Zacharias, Massachusetts Institute of Technology, Cambridge, Mass.

URUGUAY

E. Stephenson Caticha, Facultad de Ingeniería y Agrimensura, Montevideo

VENEZUELA

J. R. Almea, Oficina Central de Coordinación y Planificación de la Presidencia de la República, Palacio Blanco, Caracas
S. P. Arriecha, Universidad de Oriente, Caracas
J. F. Camero Díaz, Instituto Pedagógico, Caracas

ORGANIZATION OF AMERICAN STATES

M. Alonso, Pan American Union, Washington 6, D. C., U.S.A.

UNESCO

P. Bergvall, IBECC, Faculdade de Medicina, Av. Dr. Arnaldo, C. P. 2921, São Paulo, Brazil

(UNESCO — cont'd)

A. V. Baez, UNESCO, Place de Fontenoy, Paris 7°, France
M. C. Alvarez Querol, UNESCO Mission in Paraguay, Casilla 1141, Asunción, Paraguay

U.S. NATIONAL SCIENCE FOUNDATION

M. Hellmann, Av. Presidente Vargas 409 s/1103, Rio de Janeiro, Brazil

FORD FOUNDATION

G. F. G. Little, Av. Franklin Roosevelt 194 s/403, Rio de Janeiro, Brazil

U. S. REGIONAL SCIENCE OFFICE FOR LATIN AMERICA

H. B. Mills, Av. Pres. Vargas 409, s/1103, Rio de Janeiro, Brazil
L. M. Orman, Av. Pres. Vargas 409, s/1103, Rio de Janeiro, Brazil

UNION INTERNATIONALE D'HISTOIRE ET DE PHILOSOPHIE DES SCIENCES

F. Jaguaribe de Mattos, Rua Major Ribeiro Vaz 429, Rio de Janeiro, Brazil

MINISTERIO DA EDUCAÇÃO E CULTURA (CADES)

J. C. Mello e Souza, Av. Rio Branco 115 - 9°, Rio de Janeiro, Brazil
L. P. Mesquita Maia, Av. Rio Branco 115 - 9°, Rio de Janeiro, Brazil
A. J. de Vries, Av. Rio Branco 115 - 9°, Rio de Janeiro, Brazil
S. Markenson, R. Constante Ramos 82, Ap. 804, Rio de Janeiro, Brazil

ACADEMIA BRASILEIRA DE CIENCIAS

A. Moses, Ave. Graça Aranha 174, Rio de Janeiro, Brazil

The EMBASSY OF THE U.S.A.

A. C. Simonpietre, Av. Pres. Wilson 147, Rio de Janeiro, Brazil

The EMBASSY OF EL SALVADOR

J. M. Fishnaler, Av. N.S. de Copacabana 324, Rio de Janeiro, Brazil

The EMBASSY OF ARGENTINA

M. A. Espeche Gil, Rua Farani 29, Rio de Janeiro, Brazil

The EMBASSY OF CEYLON

A. Pathamarajah, Rua Ministro Viveiros de Castro 141, Rio de Janeiro, Brazil
V. T. Saravenapavan, Rua Ministro Viveiros de Castro 141, Rio de Janeiro, Brazil

The EMBASSY OF ISRAEL

S. Levin, Rua Paissandú 90, Rio de Janeiro, Brazil

GOVERNO DO ESTADO DE ESPÍRITO SANTO

C. L. Kulnig, Vitoria, Espírito Santo, Brazil

SECRETARIA DE EDUCAÇÃO DO ESTADO DE GOIÁS

F. Khoeler, Goiânia, Go., Brazil

SECRETARIA DE EDUCAÇÃO DO ESTADO DA GUANABARA

S. Gottschalck, Rio de Janeiro, Brazil

SECRETARIA DE EDUCAÇÃO DO ESTADO DO
RIO GRANDO DO NORTE

W. Pinheiro, Natal, RGN, Brazil

The following bodies were officially represented by participants whose addresses are given previously:

INTERNATIONAL UNION OF THEORETICAL
AND APPLIED MECHANICS

F. Cernuschi

CENTRO LATINO-AMERICANO DE FISICA

G. Fialho

THE HOLY SEE

F. X. Roser

THE NETHERLANDS GOVERNMENT

A. M. J. F. Michels

MINISTERIO DE EDUCACIÓN DE LA REPUBLICA ARGENTINA

C. A. Frumento

MINISTERIO DE EDUCACIÓN DE VENEZUELA

J. F. Camero Díaz

MINISTÈRE DE L'EDUCATION DE FRANCE

A. Payan

ACADEMIA NACIONAL DE CIENCIAS DE BOLIVIA

I. Escobar V.

THE ROYAL SOCIETY, LONDON

N. Clarke
J. L. Lewis

AMERICAN COUNCIL OF LEARNED SOCIETIES

G. Holton

COMISSÃO NACIONAL DE ENERGIA NUCLEAR, BRAZIL

F. de Assis Magalhães Gomes

COMISIÓN NACIONAL DE ENERGIA ATÓMICA, ARGENTINA

M. E. Báncora

OFICINA CENTRAL DE COORDINACIÓN Y PLANIFICACIÓN DE LA PRESIDENCIA DE LA REPUBLICA DE VENEZUELA

J. R. Almea

AMERICAN INSTITUTE OF PHYSICS

W. C. Kelly

NATIONAL RESEARCH COUNCIL OF ITALY

D. Sette

GOVERNO DO ESTADO DA PARAIBA

C. O. Lopes de Mendonça

SECRETARIA DE EDUCAÇÃO DO ESTADO DA GUANABARA, BRAZIL

E. Nunes Pereira

Program

July 1, Monday

Morning Registration
Informal gathering of participants

Afternoon Greetings from Dr. Paulo de Tarso, Minister of Education for Brazil; Professor Athos da Silveira Ramos, President of the National Research Council of Brazil; and Professor Sanborn C. Brown, President of the Commission on Physics Education of the International Union of Pure and Applied Physics

Opening Address by Professor Jayme Tiomno (Brazil), Science Education in the Contemporary World

Physics and Culture: Definition of Goals and Proposals for Science Instruction, Professor Gerald Holton (U.S.A.)

July 2, Tuesday

Morning Chairman: Professor Pierre Fleury
Secretary: Professor Francisco Magalhães Gomes

Cultural Values in Science Teaching, Professor Sanborn C. Brown (U.S.A.)

Discussion contributed by J. R. Zacharias (U.S.A.), G. A. Boutry (France), R. P. Feynman (U.S.A.), L. Bru (Spain), R. Domínguez (Honduras), E. Rogers (U.K.), N. Clarke (U.K.), F. Cernuschi (Argentina), D. Sette (Italy)

The Historical Approach in a Balanced Teaching of Physics, Professor G. A. Boutry (France)

Afternoon Chairman: Mr. Norman Clarke
Secretary: Dr. Juan J. Giambiagi

Science Education and the Humanities, Father F. X. Roser (Brazil)

Teaching Physics for Understanding in General Education: Aims, Methods, and Training of Teachers, Professor Eric M. Rogers (U.K.)

Discussion contributed by A. Harasima (Japan), F. X. Roser (Brazil), P. G. P. Leite (Brazil), F. Cernuschi (Argentina), N. Clarke (U.K.)

July 3, Wednesday
Morning   Chairman: Professor Oscar Sala

Curriculum Reform in the U.S.A., Professor Jerrold R. Zacharias (U.S.A.)

Discussion contributed by F. X. Roser (Brazil), P. Lorrain (Canada), E. Rogers (U.K.), P. Fleury (France), J. S. Pupo (Brazil), J. Tiomno (Brazil), I. Raw (Brazil), E. Ingelstam (Sweden), F. Cernuschi (Argentina), R. Canhato (Brazil), R. Domínguez (Honduras)

Afternoon   FREE

July 4, Thursday
Morning   Chairman: Professor M. Valouch

The Aims of Elementary and Secondary School Science, Professor A. M. J. F. Michels (The Netherlands)

Science in Elementary and Secondary School Education, Mrs. C. A. Michels-Veraart (The Netherlands)

Afternoon   Chairman: Professor D. Sette
              Secretary: Professor Onofre Rojo

Principles of Classroom Demonstrations and Laboratory Work in Teaching Physics, with Examples from Recent Swedish Developments, Professor Erik Ingelstam and Dr. Gustav Friskopp (Sweden)

Discussion contributed by J. S. Pupo (Brazil), J. Lebrón (Paraguay), D. Sette (Italy), S. Markenzon (Brazil), F. Cernuschi (Argentina), Alvarez Querol (UNESCO), J. R. Zacharias (U.S.A.), P. Fleury (France), J. O. Reyes (Dominican Republic), G. Holton (U.S.A.), D. Golman (Brazil), J. Tiomno (Brazil), D. Moreno (Chile)

July 5, Friday
Morning   Chairman: Professor A. M. J. F. Michels
              Secretary: Dr. Ismael Escobar

New Experiments in Teaching Physics, Professor D. Sette (Italy)

Physics Teaching in Czechoslovakia, Professor M.
Valouch (Czechoslovakia)

Discussion contributed by M. Nicolau (Mexico), M.
Báncora (Argentina), F. J. Mattos (Brazil)

The Place of Atomic Physics in General Education,
Mr. John Lewis (U.K.)

Discussion contributed by M. Báncora (Argentina),
D. Sette (Italy)

Afternoon   Chairman: Professor Paulus A. Pompéia
Secretary: Professor Gabriel Fialho

The Teaching of Physics in Underdeveloped Countries, Dr. Albert Baez (UNESCO)

Discussion contributed by L. Bru (Spain), J. R.
Zacharias (U.S.A.), F. Cernuschi (Argentina), G.
Fialho (Brazil), D. Sette (Italy), P. Fleury (France),
G. Holton (U.S.A.), J. Tiomno (Brazil), P. A.
Pompéia (Brazil).

July 6, Saturday
Morning   Chairman: Professor Sanborn C. Brown
Secretary: Dr. Isaias Raw

Summary of the Conference, Professor Sanborn C.
Brown (U.S.A.)

Discussion contributed by J. R. Zacharias (U.S.A.),
G. A. Boutry (France), A. Bueno (Peru), J. Herkrath
(Columbia), J. S. Pupo (Brazil), P. Fleury (France).

Summary
of the Conference

This Conference was held to discuss some particular aspects of general education. By general education we have meant the broad training to be given to all children up to about the age of sixteen, a training that is not intended to prepare them for any particular occupation but which is the best preparation that we can devise for their entering into the adult world. It is inconceivable that in the modern technological age it could be seriously disputed that science should be an essential part of this education. Physics is the most fundamental of the sciences and therefore is clearly essential. Our view does not rest upon the obvious usefulness of physics in a technological world. On the contrary, we have stressed that the claim of physics to be part of the education of all children stems from the place that the subject occupies in the intellectual heritage to which we seek through education to introduce our children. Moreover, it is for this very reason that we are agreed that physics courses commonly given to children are inadequate and unsatisfactory. They fail to present in any convincing way the magnitude and grandeur of the intellectual achievements represented by modern physical theory, they fail to convey understanding of the more important concepts of physics, and they fail to show to young people anything of the approach which the professional physicist necessarily brings to his work.

We have discussed a number of actual developments in physics courses in various countries, and these we have welcomed. No single course will meet the needs of all countries with their varied educational systems and their different problems of finance and teacher supply.

We strongly recommend that in every country both educationalists and governments should be acquainted with the important work in this field currently being done in Europe, the U.S.A., and elsewhere, and should seek to use such of this work as may best be adapted to their own needs. We are not narrowly concerned with our own subject, which, in the context of general education, we recognize as but one facet of human culture and scholarship. Indeed, we have discussed the necessity to teach physics in such a way as to ensure an understanding of the many opportunities that it offers of links with other fields of learning. Since a very small proportion of the population of any country will be professional physicists, we have discussed how to teach science so as to make it a working tool in the life of the educated man. To do this, it is essential to make physics interesting so that children will want to learn. It is also vitally important to teach the conceptual framework of

1

physics, which can be remembered throughout life, and not merely individual facts, which are easily forgotten.

We are unanimously agreed that physics must be introduced at an early age and must be taught throughout the education of the child. We have discussed for many hours how to accomplish these objectives in specific ways.

Finally, we discussed how to teach our teachers so that they themselves can understand the cultural nature of physics. Throughout there was unanimity that one must concentrate on a few subjects as illustrative of the scientific approach, and that it would be a grave error not to teach some branches of physics in depth. Physics must be taught as physics, even though one is teaching those who will not become professional physicists. It is impossible to learn physics without going deeply into the subject.

Various schemes were discussed whereby mere shallow surveys of physics could be avoided and selected branches taught in depth while presenting a good broad picture of the subject.

Specific schemes discussed in detail were the PSSC, the Nuffield Foundation Physics Project, which is in a way similar to PSSC but designed to cover several years rather than one or one and a half. We discussed the pilot project that UNESCO is sponsoring with IBECC (the Brazilian Institute of Education, Science and Culture), and, under this heading, also the value of programmed instruction. We heard about many individual experiments, one of the most interesting being that of the Mobile Units in Sicily, which are applicable to countries which do not have an adequate supply of properly trained teachers or teaching apparatus.

We have had long discussions on the place of films. If it is not possible for students to have laboratory facilities or for an experiment to be done in the lecture, either because of lack of apparatus or for other reasons, films have great educational value, but a film should be made so well that it demonstrates the general philosophy in the film itself.

We also had an apparatus exhibit set up by various organizations to illustrate simple and available equipment for the teaching of physics.

Opening Session

At the opening session of the Conference, welcoming addresses were given by Dr. Paulo de Tarso (Minister of Education for Brazil) and Professor Athos da Silveira Ramos (President of the National Research Council of Brazil). Ambassador Paulo Carneiro, who had hoped to be able to address the Conference, was unfortunately unable to attend. The first plenary session then opened with an address by Professor Jayme Tiomno of Brazil.

The Conference was addressed by His Excellency, The Minister of Education for Brazil, Dr. Paulo de Tarso, as follows:

I have the honor to welcome you on behalf of the Brazilian Government and to convey to you the personal greetings of His Excellency President João Goulart. I must say that for Brazil it is a privilege to be able to receive the visit of such eminent scientists and such experienced educators, intent on making the most of these debates on physics in general education and upon the training of the scientists and the technicians that the present-day world and our country in particular so urgently need.

I can but tell our renowned visitors and participants in this Conference that we are a people engaged in a bitter struggle against misery. We are determined to conquer misery; we are confident of the success of our struggle against underdevelopment.

We refuse to accept misery as a predestined affliction; to the extent that you scientists succeed in enlarging the field of achievement of the human mind, victory over misery is brought nearer. And what used to be considered inevitable is scaled-down to the proportions of an iniquitous fact that can and must be put right. As a people of peaceful and peace-loving traditions, the Brazilians are anxious, through me, to send you fraternal greetings and, in all humility toward the participants in this Conference, to express their confidence in physics, more particularly, and in science in general, not as an instrument for the domination of some men over others, but as a means whereby man is enabled to dominate nature. While it is true that in wartime the need to destroy has stimulated scientific discovery, it is, on the other hand, hopeful to observe that now in peacetime, understanding between men and brotherly love have also served as efficient motives to urge science on to new and meaningful achievements. There were those who alleged that manipulation of the atom was tantamount to giving man an option on himself and providing an opportunity for the mass suicide of humanity. It is necessary for human progress to keep pace with scientific progress, which, thanks to men like you, has been so significant, if the atom is not to be a weapon for the self-immolation of mankind but rather the means of promoting the moral and material advancement of peoples as a whole. This is the fundamental significance of the message that I want to convey to you on behalf of the Government of my country, to which I have the honor to belong.

But before concluding, I should like to sum up for you my feelings as Minister of Education and Culture of Brazil, feelings motivated by the special appreciation and deep satisfaction of having so important a Conference held in our midst. These ministerial premises, this House as we would say in Portuguese, must be the house of the New Science that rends the veils of space, plumbs the mysteries of handling the atom as venturously as it explores the labyrinthine depths of the subconscious, and is, in fact, the Science of man, resurgent in this second half of the twentieth century and extending his power continually over Nature so as to transform her wild energies through science into culture for Mankind. Education should be at the service of science, for science is engendering the prospects of culture. Education, science, and culture thus emerge as three phases in the cyclic process of man's self-fulfillment.

Vast is the yawning abyss between the ability with which science is leaping ahead and the stagnant situation in the underdeveloped countries where the human groups that form the majority live on the barren fringe of education, science, and culture.

If we want science to serve mankind and not to impose upon us the blind technocracy of a world of robots, if we want the riches that flow from scientific discoveries to be absorbed into the cultural legacy of humanity, then we must fling open the portals of knowlege to the multitudes that are thirsting for knowledge and eager to discover themselves. In this endeavor we must work together as citizens of the world and bend all our efforts to the furtherance of the cause of humanity.

Professor Athos da Silveira Ramos, President of the National Research Council of Brazil, also spoke.

We have the privilege of living in the very midst of the Decade of Development, according to the conclusions of the General Assembly of the United Nations meeting in 1961. And for this very reason our responsibilities as scientists are made all the heavier, inasmuch as the scientific discoveries totaling a figure never previously attained, of some three million original scientific studies a year, accomplished by about two million active scientists, embody or imply technological consequences that may improve living conditions in the less developed countries but may also increase even more drastically the wide gap, measured in terms of social welfare, that separates the rich countries from the poor ones. Every day, the frontiers of knowledge are thrust forward and are likely to reach unforeseeable bounds in the near future, while at the same time technology finds practical applications for science with amazing rapidity. The physicists of the twentieth century have given mankind the necessary elements in sufficient quantities to disintegrate matter and recover fabulous stores of energy; to materialize energy with a no-less-spectacular energy yield; to overcome the force of gravity and explore the unknown realms of outer space, with all its promise of worthwhile discovery; and in so doing, they have concentrated in physics the most alarming fears of self-destruction and the most encouraging hopes of better days for this humanity of ours, so uniform in its weaknesses, but so diversified in the vigor or fragility of its scientific knowledge, and in the living conditions imposed by development or underdevelopment, as the case may be.

Scientific education is the only road to progress, and physicists are well aware of it, lending the prestige of their presence to this outstanding occasion for raising culture to a universal plane, which is opening today with the valuable collaboration of such high authorities as the Minister of Education and Culture and scientists of such brilliant and far-reaching renown. Modern man cannot do without a certain leaven of science in his general education to enable him to discover what he really is, identifying the potential scientist in his make-up, which is what amounts, at the present time and for any country, to the most significant assurance of progress and social welfare. At this moment when physicists all over the world are striving to imbue with a basic knowledge of their science the youth of today that will be responsible for the world of tomorrow, I should like to make known — possibly for the first time in the Portuguese wording — the tenor of the Geneva Declaration, made by the members of the International Scientific Community meeting there, which was the outcome of the successful persuasion exerted by Brazilian Ambassador Josué de Castro on the members of the First United Nations Conference on the application of the sciences and technology to underdeveloped areas and bore the signatures of the 159 representatives of 107 participating countries. The Declaration (translated from the Portuguese) runs as follows:

"We, the undersigned, members of the International Scientific Community meeting in Geneva to take part in the United Nations Conference on the application of science and technology for the benefit of less developed areas:

"Aware that in spite of the extraordinary progress of science and technology made in the last few decades, about two thirds of humanity are still afflicted with hunger or malnutrition, avoidable endemics, ignorance, and want;

"Considering that a great international effort will be required to overcome these conditions, and that such an effort demands an unprecedented mobilization of modern sciences and technology to support development of less favored areas:

"Recognizing that it would be easier to carry out these intentions with a wider range of international cooperation and a considerable expansion of the present scale of aid;

"Considering that the capital available for this purpose is seriously curtailed by the huge, unproductive world outlay on armaments;

"Recognizing that the problems of disarmament and economic and social development are the two major challenges that mankind will have to face in the years to come;

"Inspired by the resolution recommending the peaceful use of such funds as may be set free by disarmament, passed by the XVII General Assembly of the United Nations in December, 1962;

"Desire, in our individual status as members of the International Scientific Community, to express our belief that all the peoples should focus their greatest possible endeavors on achieving the goal of complete general disarmament, and seize the opportunity offered by the resumption of the 18-nation Conference on Disarmament, meeting in the city of Geneva, to urge the conclusion of an agreement on the suspension of experiments with atomic weapons and the achievement as soon as possible of complete general disarmament under effective international control, thus liberating the funds desperately needed for the great constructive task of promoting the social and economic development of the peaceful world."

As may happily be observed, the scientist living in the twentieth century, forming as in the past the minority to which the rest of mankind owes its scientific progress, for the first time in history is trying to sway international politics toward the recognition that the discoveries of man should be placed at the service of mankind as a whole. Convinced and moved by the depth of this scientific and social declaration and its postulates, I address the most cordial and enthusiastic greeting of the National Research Council to the eminent scientists who are now visiting us, according us the honor of personal acquaintance and bringing us up to date with respect to the latest evolutionary trends of science, the grandeur of which knows no frontiers and never will.

1

Physics Teaching in an Underdeveloped Country

SCIENCE EDUCATION IN THE CONTEMPORARY WORLD*

Professor Jayme Tiomno (Brazil)

It is nowadays generally agreed that the explosive development of science in this century is one of the major events in the history of human culture. The first half of this century stands out as a historic landmark, as notable as that of the age of great inventions which signified the transition from the Middle Ages to the Modern Age. The ease of communication and of the transmission of ideas, the actual range of scientific, technological, and industrial development, and the expansion of educational systems in the last century have, moreover, greatly augmented the impact of recent advances and their more immediate consequences. Hence, although we cannot accurately assess the far-reaching changes through which mankind will pass in the next few years, there is no doubt that these will be enormous. We even feel the need to give a new name to the present stage of history; the era of interplanetary flight, the era of atomic energy, the era of industrial automation, and many others have been given. These seek to associate our age with the most spectacular successes of modern technology, which in turn arise from scientific progress. The first of these descriptions, Interplanetary Age, emerged immediately after the launching of the Sputnik, which was an achievement that aroused widespread emotion among the masses of mankind. This was probably because outer space and the heavens have hitherto been associated with religious sentiments and have been considered inaccessible to man. This fact and others, as well as the recognition of the steady increase in the conditions of comfort created by the advance of technology in the more highly developed countries,

---

*Material presented by the author has been subject to extensive editorial revision.

contrasting with increasing difficulties in underdeveloped countries where this technology is lacking, tend to generate the conviction that scientific development is indispensable to economic development. This attitude of the man in the street toward science is new and peculiar to this century. It is contrary to the traditional tendency to regard science as an esoteric or cabalistic activity, and scientists as strangely abnormal persons engaged in mental processes that are incomprehensible to the common run of mortals and are wholly remote from everyday experience. The growing interest and curiosity of the common man in scientific subjects has increased their cultural importance, and appreciation is spreading to ever wider sections of the community. It is clear that, while the recognition of science may be penetrating increasingly into the minds of the masses, its more technical and skilled branches, which require a grasp of detail and a capacity for performance, must be restricted to those who have been specially and carefully trained for the purpose. The importance of science in general education and in shaping the mind of the ordinary citizen is being increasingly recognized, except perhaps in the most backward countries; the belief that science is opposed to the humanities has been left far behind. The progressive, rapid change from an educational system designed to produce an intellectual elite to a system of universal education, as well as the need to educate technicians and scientists on a widening scale, forces us to admit the failure of the teaching systems in use during the first half of this century. The development of educational methods more firmly grounded on a substratum of scientific knowledge, psychology, and other sciences, has made it possible to tackle more systematically the problems of scientific training for the average citizen and for the specialist, in order more efficiently to meet the requirements of the modern world. Recognition of the need for drastic revision and reform in the methods and systems of scientific education and in physics, in particular, has instigated experiments in various countries, the results of which are worthwhile comparing. It is interesting to note that most of these experiments were made with the participation of leading scientists who were the first to realize that education had to be revised. The attendance, at this and other Conferences on the teaching of physics, of outstanding researchers is a proof that the modern scientist no longer lives in his legendary ivory tower. On the contrary, he takes as active an interest as the average citizen, or an even more active one in the problems of his national community and in problems of international significance. In the present case, the responsibility of scientists and educators is particularly heavy. They are the most able persons to point to solutions and to bring home to governments and to national and international institutions the need for planning and for taking urgent steps to ward off future disasters and the hazards of irreparable damage to the development of their countries.

Of course, local conditions vary considerably from developed to underdeveloped countries, but the general approach to the problems of teaching physics is essentially the same, though the solutions may differ in accordance with the differences of local conditions. Developed countries have long gone beyond the stage of primary education for all, and are overcoming the difficulties in the way of universal secondary education; they are now well on the way to university education for everyone. The fundamental problem there is how to teach science to a large number of students and with greater efficiency, the object being to give an adequate scientific background to nonspecialists and to train specialists in sufficient quantity and quality to meet the requirements of the nation's scientific, technological, and economic development.

It is a propitious sign for humanity as a whole that the cold war between the world's two greatest powers is being replaced by technological and educational competition. At the basis of these ventures are facts such as the effort these countries are making to turn out engineers and scientists to a total of about a hundred thousand a year. This certainly has tremendous implications for the teaching of physics at all levels and calls for a thorough overhaul of the educational process. In the underdeveloped countries, the inadequacies of the educational systems may be traced to various historical conditions, including only recent freedom from foreign rule. Often there are not enough teachers, and far fewer scientists and specialists. In countries where independence has been enjoyed for many years, although there are competent teachers, specialists, and even scientists of international standing, their numbers are insufficient to cope with the increasing needs. In such countries, generally speaking, the government is only now awakening, under the urge of public opinion, to the glaring demand for students to be admitted in larger and larger numbers to education at all levels. Often enough, the solution adopted is merely that of admitting more students to the schools without tackling the problems of increasing the number of teachers and researchers and the reorganization of teaching methods. An attack on these probelms cannot, however, be further postponed without increasing the relative degree of underdevelopment of the country. The systems mentioned cover a wide range of gradation; however, the attendance at this Conference of representatives from countries at the most widely varying stages of development points to the existence of a common field of interest and even of general principles, to form a starting point from which discussion can be engaged on specific problems and from which experiences may be compared.

Finally, I want to emphasize that the challenge that the world of today, with its content transformation, is presenting to all nations is terrible. Each must either keep up with its rate of development or perish. And it is by investing on an increasing scale in scientific development, in the training of competent specialists and teachers, and in expanding the educational system and bringing it up to date that the nations can fit themselves to stay abreast of the tide of progress in the modern world, and so survive.

The problems and difficulties of physics education in Latin America had been extensively discussed in the First Inter-American Conference on Physics Education (Rio de Janeiro, June 24-29, 1963). The recommendations of that Conference are helpful in showing the shortcomings and difficulties in that area. Some of these are clearly presented by the following summarized resolutions:

SUMMARY OF RESOLUTIONS ADOPTED BY THE

FIRST INTER-AMERICAN CONFERENCE ON PHYSICS EDUCATION

- The teaching of physics at all levels should be based primarily on experimental work instead of on the mere accumulation of information.
- The teaching of physics in high schools should be adapted to the intellectual level of the children at this stage.
- Strong support is needed from national governments and international organizations to develop programs for the production of auxiliary teaching tools, particularly inexpensive equipment, and of films in the national language of the country.
- The training of physics teachers should be carried out at universities or at special institutions of university level (and not at institutions that are at the secondary-school level, except for the practical teaching of pedagogic subjects).
- In training physics teachers, emphasis should be given to scientific education in physics and an adequate knowledge of chemistry and mathematics.
- National agencies that grant scholarships should have advisory bodies which include scientists and university professors.
- Efforts should be made to introduce generally the system of full-time appointments for university staff, as well as the benefits of the sabbatical year of leave.
- University teaching of physics, even for technical careers, should be the responsibility of physicists.
- Existing plural physics departments at the same university (often up to six) should be amalgamated.

Many of the Latin American participants emphasized the unsatisfactory state of the teaching of physics in underdeveloped countries. Typical of their point of view are the comments from a paper submitted by Professor P. G. de Paula Leite.

OBSERVATIONS ON THE TEACHING OF PHYSICS

IN DEVELOPING COUNTRIES*

Professor P. G. de Paula Leite (Brazil)

I am aware of the dangers of generalization. My experience being mainly restricted to Brazil may lead to conclusions that are not applicable to all other developing countries.

Let me stress some points that are very unsatisfactory in the teaching of physics in underdeveloped countries, both at secondary-school and at university level.

As a general picture, teachers are almost only concerned with their task of instructing rather with problems of education. They pay much more attention to the description of facts and apparatus and to the formal handling of equations than to developing in students the ability to use physics and to understand physical phenomena and concepts.

The most common failures of students who finish secondary schools in underdeveloped countries are the following:

1. Lack of ability to do experimental work even of a very simple nature, and a distaste for doing any manual work.
2. Difficulty in reporting on the results of their observations.
3. Difficulty in working by themselves on situations that were not discussed in detail during the lectures.
4. Lack of interest in any subject that was not worked out in the classroom, and even a marked difficulty of studying alone with textbooks that do not follow the same approach as that of their lectures.
5. Lack of adaptation to teamwork.

These shortcomings are the consequences of the teaching methods adopted in the high schools and of the fact that the children do not have much direct contact with technical developments and gadgets in their daily experience.

As far as the actual teaching is concerned, there is a lack of training in educational psychology in the presentation of the subject according to the intellectual development of the children.

Another point which is worth mentioning is that the teaching of physics in secondary school is mainly directed toward preparation for entrance examinations at the university, mainly for technical careers. The

---

*Material presented by the author has been subject to extensive editorial revision. This paper was submitted to the Conference but not formally presented.

shortage of places at the university and the inappropriate type of examination have terrible consequences. Physics teachers in high school ignore the need for children to understand fundamental concepts and be familiar with experiments. They concentrate their attention on the solution of problems of the kind that are set in the examinations. Thus the students do not really learn any physics but concentrate on cramming for the strenuous competition of the entrance examinations. No improvement can be achieved until these examinations have been drastically changed. Then a completely different approach to the teaching of physics, somewhat on the lines of the PSSC program, might be attained.

Physics teaching at the university level presents no brighter picture. Having to supplement the wholly inadequate foundation given in the secondary school and having even to eliminate erroneous concepts from the students' minds, the teachers fail again by insisting on any encyclopaedic coverage. The lack of experimental equipment, the excess of students, and the shortage of instructors make the situation worse. The pressure to restrict the teaching of basic sciences to the first year so that immediate professional training can be started also has its ill effects. As the students have no time to think or to understand, they cannot enjoy the benefit of a solid foundation of basic science.

I should like to propose that, in order to improve the teaching of physics in developing countries, we should act as follows:

1. Modify completely the entrance examinations at the universities, so that they find out how much physics the student really understands.

2. Adapt the PSSC program to local conditions and train high-school teachers in these methods.

3. Set up national committees for the teaching of physics and require university professors and high-school teachers jointly to study ways of rapidly improving the present situation.

2

Science as a
Part of Culture

CULTURAL VALUES IN SCIENCE TEACHING*

Professor Sanborn C. Brown (U.S.A.)

Ours is an age of explosive scientific expansion. The number of people employed in scientific pursuits is growing so rapidly that all other elements of our culture can be said to be standing still by comparison. Statistics show that the number of people employed in non-scientific intellectual pursuits doubles every forty to fifty years; in science the number doubles approximately every ten years.

This tremendous and rapid growth of science has resulted in a phenomenal involvement of our culture with science, but there has not been a corresponding acceptance of science as part of our culture. As physicists we are all aware of this. We are constantly confronted with evidence that the world at large looks on our field as one reserved for the narrow, mathematically endowed specialist, contributing only to our technological surroundings, talking a special language of his own, occasionally nowadays honored for his intellectual achievements that, however, the public makes no effort to understand.

How often do we hear intelligent leaders of our society almost boasting that they never understood physics. The same people would blush with shame if they could not discourse with depth and understanding on art, literature, and philosophy. Why is this? Why, in an increasingly technologically based society, does not that society recognize the cultural values of the science upon which it is coming more and more to rest?

Part of the trouble lies with a major misconception in the mind of the public concerning what science is. The layman thinks of science most often in the context of what might be called the "Conquest of Nature." By this he means the spectacular achievements of technology and medicine, nuclear power, polio vaccines, and the like. He does not think of science within the framework of concepts that have an important

---

*Material presented by the author has been subject to extensive editorial revision.

impact on our thinking and beliefs. How often do we discover that the nonscientist thinks of science in terms of just "facts" derived from empirical investigations, or in terms of rigid laws and far-reaching, but inhuman generalizations. How seldom do we find intellectual leaders in the humanities recognizing that contemporary science is essentially a symbol system which enables men to develop postulates and assumptions with which to observe, perceive, and interpret events. Nor have we yet persuaded the nonscientists that we recognize that the scientist-observer is himself in the picture and that whatever he observes and interprets is patterned by his own basic conceptions and criteria of credibility.

I believe that the reason for the public's image of science in general and physics in particular derives from the attitude of our schools. Our whole approach to science teaching has been wrong. We have been teaching physics as a great catalogue of facts. We have presented physics in the framework of laws to be learned, of formulae to be brought out and applied to solve problems, and of routine laboratory exercises aimed at arriving at predetermined answers.

We, here, are all aware that the growth of knowledge in science has been so explosive that basic changes in educational philosophy have been forced onto us, and that these changes must be guided not only by the expansion of knowledge but by a real desire to make every educated man in the coming generation appreciate that science is a basic element in our culture. The scientist and the nonscientist must learn to understand each other and to accept common goals and values.

I believe that the directions in which we should go are clear. The philosophy of education should reject the notion that there are a large number of specific items of information to be learned, or that extension of knowledge occurs merely by the simple process of adding facts. Any pretense of covering a field of knowledge is obviously ridiculous and must be abandoned. Rather, reliance must be placed on the apprehension of a system of basic concepts and their logical consequences. The material to be included in a particular course of study should be selected not so as to identify particularly worthy parts of knowledge but to identify the concepts that would be most fruitful in advancing understanding. The content of a course should be arranged so as to enable the student to proceed from the initial grasp of a concept to a more comprehensive understanding, and to see the interrelationships of the basic concepts.

The view that I have just expressed tries to take into account not only the rapid extension of knowledge which is occurring but also the fact that knowledge is constantly being reorganized and reconstituted. Our only real hope of keeping up with rapidly advancing knowledge is to develop the ability to rethink what we have already learned rather than merely to remember what we already know. Thus the process of education becomes not so much a process of adding to one's information as a process of acquiring a number of different methods of inquiry through which we may define learning goals, give order and rigor to learning activities, collect data and analyze them, draw inferences and test them, correct errors in theoretical formulations, and reformulate hypotheses to accommodate new data.

Closely associated with this concentration on the understanding of basic concepts and their logical consquences is the recognition that man's ability to learn facts and to remember them deteriorates rapidly with age. Thus, the facts of science which we today feel to be essential to our understanding of modern physics may well be unimportant twenty years from now. The real importance of this is, of course, that our students of today who will be the intellectual, industrial, and political leaders twenty years from now may well find no use for the facts of physics of today but will be faced with a new and different physics. If we have not equipped them with the basic conceptual framework within which to re-examine and reformulate the physics they are now learning, they may well reject science just at the time it is most important that they recognize and use it as a basic element in the culture of the society they are called upon to mold. We must be teaching today for the boundary conditions of the future, and only insofar as we concentrate on instilling into our students the understanding of a system of basic concepts and their logical consequences can we hope to ensure that science can take its place as a fundamental element of our cultural values.

So far I have concentrated on the specifics of science education. I am sure it is obvious to all of us that this is a reversible process. The cultural value of teaching science is as much the responsibility of the teacher of humanities as it is of the teacher of physics. The task of education today is not to teach "the best that has been known and thought in the past," but more and more it must be to orientate students so that as adults they will participate in the variety of roles and activities that are necessary for their leadership in society. If the humanities are to live up to their responsibilities in modern education, they too must modernize to the extent that their love of the past does not deprive students of learning about the present as the matrix for the culture of the future.

As science and technology are playing a greater and greater role in society, we recognize more and more how frequently the teachers of humanities have lacked not only a concern for but even an awareness of the role and function of science. Yet they too have a responsibility for providing students with an awareness of modern society based on the fruits of science. Humanization of knowledge can be said to be the process of communicating basic concepts and assumptions and of presenting a variety of models for direct experience, whereby a student learns to orientate himself to what he is learning. If education is to be more than the training of future scholars, scientists, and professional men, what we teach must be so humanized that students will be helped to live both today and in the future, neither completely ignorant of the world of today nor completely immersed in the past. These goals cannot be accomplished without a full recognition of the emergence of science as a basic element in our culture in the present and to a greater and greater degree in the future.

How then should we focus our teaching so that science is a basic element in our culture? Are there basic changes in our approaches which would make more obvious the cultural values of science?

As a start in answering these questions, let me draw on a paper by Professor Michels[1] from the Netherlands. He points out that the activities of education can be divided into three parts: (1) The acquiring of knowledge and facts, (2) the application of this knowledge to the task at hand, and (3) the deep penetration into the fundamentals which produces a basic understanding of the interrelationships of the knowledge and facts that may lead to aesthetic as well as to the philosophic implications of this knowledge.

It is in Phase 3 that the cultural values of science lie just as in any other discipline. The failure of the general public to recognize this cultural value arises from the fact that scientists have emphasized the first two pragmatic phases to such an extent that the third, and perhaps most important, phase has been lost by comparison.

Surely, however, these three phases are typical of many branches of human endeavor. Let me draw on two illustrations from the humanities. Take, for example, the study of language. Phase 1 consists of learning the words and grammar; Phase 2, the application of this learning to reading and writing. Here we have the tools for communication and for acquiring further knowledge. But the real essence of the cultural values of language does not come until Phase 3, where prose and poetry are brought to bear on the human character — man's hopes, his aspirations, his loves, his hates, and the whole gamut of his emotions.

To press home my point let me take another example, this time from the field of art. In Phase 1, the student must learn the use of materials and media — the paint, the brush, the chisel, the canvas, the metal, or the stone. In Phase 2, he learns to form the drawings, to put the paints together, to express his art form in a unified whole. However, we do not recognize Phase 2 as real art. It is not until human emotions are expressed, not until human feelings of the artist himself are transferred to the canvas or the bronze that we reach Phase 3, and something of real cultural value has been contributed.

Surely science has the same three phases. Phase 1 contains the collection of facts, the laws and postulates, the mathematical models and formulations, the array of basic building blocks that so frighten the nonscientists. Phase 2 involves the application of this knowledge to the extension of knowledge and to the technologically useful devices that the layman so often confuses with science itself. We here are all physicists, and we understand Phase 3, the appreciation of the understanding of nature, its unity and its beauty as well as its impact on the lives and emotions of modern man. Why is it that we are such a small group in understanding this?

I believe that the answer to this question lies in the way we have been teaching science for the past few generations. We have been teaching only Phase 1 and Phase 2. We have been teaching the collection of facts, the laws, the postulates, the mathematical models, and formulations and bragging about the extension of knowledge and the usefulness of our advancing technology, but nowhere have we put man into the system. Phase 3, which encompasses our real contribution to human culture,

---

[1]Private communication.

contains the interrelationships of scientific knowledge to man, not only as an intellectual being but as one with aesthetic, emotional, and philosophic values as well. Here, it seems to me, science has not yet made its position clear.

Let us think for a moment about those subjects which we usually call "cultural" subjects: the study of language, literature, art, history, philosophy, and many more; and let us analyze how much time teachers in these subjects spend in Phases 1 and 2. The answer will obviously be that, to a first approximation, all of their time is spent in acquiring knowledge and facts and applying this knowledge to the details of the discipline under study. Seldom does a teacher of language, literature, or history reach Phase 3 — the deep penetration into the fundamentals which produce a basic understanding of the interrelationships of knowledge in the particular field and basic human values. If Phase 3 is reached so seldom in those disciplines that are called the "humanities," what is different about our sciences or our teaching of the sciences which puts them, usually, outside the realm of "cultural" subjects?

Once again, I believe the answer lies in the way we have been teaching science, not so much in our colleges and universities but to our children. The image we build up of science and its relationship to man's cultural surroundings from the very earliest years of elementary and secondary school is that of an intellectual robot. If by the time our young men and women reach the university they are already persuaded that there is a basic cleavage between the sciences and the humanities, we have already failed in our teaching task.

You will notice here that I am somewhat at variance with Professor Holton[2] in the brilliant address he gave to us. I am very much in favor of Professor Holton's proposals at the university level, but I think they should come much earlier in the intellectual development of our students. It is true that at the present time our elementary- and secondary-school science education is so sterile that it is at the university that we must try to persuade students that science can be integrated into our culture. For this reason I am in favor of Holton's suggestion: I hope that we can so improve our preuniversity education in the coming decade that Professor Holton's constellation scheme will seem obvious. That not only physics courses be intimately connected into the whole intellectual development of the educated man, but the humanities themselves will incorporate the thinking of science into their teaching of contemporary culture.

To accomplish the goal of incorporating the cultural values of science into our over-all educational framework, my plea is that we start as early as possible in the intellectual development of our children. I am impressed by the hypothesis so ably championed by Jerome Bruner[3] that any subject can be taught effectively in an intellectually honest form to any child at any stage of development. Let us concentrate on this insight, and let us develop an elementary and secondary-school

---

[2]See page 27.
[3]Jerome S. Bruner, Progress of Education, Harvard University Press, Cambridge, Mass., 1963.

teaching of science which rejects the notion that there is a large number of specific items of information to be learned and concentrate on the apprehension of a system of basic concepts and their logical consequences.

Man today can rarely think or act independently of the influence of science. Let us so communicate the essential implications of our methods and findings that it is a basic element in every child's educational development. Then and only then can it really be said that the values of science have become a real element in our culture.

A lively discussion followed Professor Brown's address. Professor Zacharias (U.S.A.) questioned Professor Brown's reference to the work of Bruner on the theory of learning. Bruner's views were valid only for subjects like mathematics, and it was true that complicated ideas could be taught in terms of very simple mathematics. It was not, however, possible to generalize and to say that if we understood any subject sufficiently well ourselves, we could teach it to any child.

In reply, Professor Brown pointed out that he had neither stated nor implied that such a wide generalization could be made. He had, however, had personal experience of teaching small children and he was satisfied that, by the careful choice of language, it was possible to give to even small children a satisfactory grasp of those concepts of physics which fitted into the children's background knowledge of fact.

Professor Feynman (U.S.A.) expressed fundamental disagreement with Professor Brown and took an extreme point of view, which was not echoed by any other speaker throughout the Conference. He questioned whether anyone yet knew enough about teaching physics to nonspecialists to justify discussing the subject on an international basis. Moreover he believed that "there is a difference between a science and the humanities and an attempt to mix the two together at too early an age is a danger and a destroyer of the true cultural value of science." The development of scientific investigation had been a revolution in culture. Science required clear thinking, a knowledge of one's hypothesis, and constant reference to experiment which was a guide to truth that was independent of authority or of opinion. These were "the main characteristics which distinguish science from other cultures and which represent, from some points of view at least, the true culture of science. Therefore one may argue that science should be taught in the purest way possible." Furthermore, "It is impossible to teach appreciation of anything to young children; you can teach them only what the thing really is and then hope that the intelligence will produce the appreciation."

Mr. Norman Clarke (U.K.), in contrast to Professor Feynman, stressed that the problems discussed by Professor Brown were most practical and, indeed, very urgent. In the United Kingdom most of the children in grammar schools, about one quarter of the total number of children, were taught science for four or five years. Of those grammar school children, no more than one in ten would eventually do anything in life which required that knowledge of science. Insofar as science was essential to the economic development of countries, which was certainly a very important factor, the problems of scientific education concerned only the 10 per cent who would later use it. What then was the justification for teaching science to other children? Surely it was important to create among educated people a general sympathy for science and an understanding, in very broad terms, of science and the activities of professional scientists. It was necessary to dispel the impression that a scientist was someone who worked in a specialized and semimagical way in a subject which ordinary people could know nothing about. To achieve these objectives and to justify, therefore, the teaching of science to those who were never going to use it professionally, it was necessary to devise courses that presented in intelligible terms a broad picture of the essential parts of science, together with an indication of the links between science and other fields of culture. Unless this could be done, there was no purpose in teaching science widely, and those countries that are at present doing so were wasting the time of thousands of teachers, the time of millions of children, and valuable material resources.

Professor D. Sette (Italy) believed that science must be part of the intellectual equipment of every man and that in education the unity of culture should be stressed. It was necessary to have various schemes such as the one proposed by Professor Holton as a transition from the present unsatisfactory situation to some other system of education that we had not yet been able to see clearly. Referring to Professor Zacharias' comments, he pointed out that one should not interpret Professor Brown's remarks on the work of Bruner as meaning that we wished to teach relativity to a six-year-old boy. It was always possible to chose suitable scientific subjects that could be well taught and illustrated to young people of any age and state of development so as to convey the essence of science, which is the continuous interplay between experiment, logical deduction, induction, and hypotheses.

SCIENCE EDUCATION AND THE HUMANITIES*

Father F. X. Roser (Brazil)

The vigor and speed of the scientific revolution that we are observing today and the apparent paralysis and decadence of the humanistic methods of yore are producing a generation with a mentality which might be considered somewhat restricted, not really educated.

Extension of knowledge increases fast day by day. This adds to the burden of the expanding student population that has to absorb such knowledge rapidly.

The classic humanistic education which prevailed in the past has been succeeded by technical and scientific education which is making itself essential to the present age. It would appear that there is a radical opposition between the two approaches. On the one hand, there is the static attitude of the humanistic training, following the great philosophical principles, ethical concepts, and historical ideas about the universe, and considered as a kind of privilege of each cultured individual. On the other hand, there is the dynamic, inductive culture based upon the application of experimental techniques to common life, represented by technical and scientific training. But, in fact, these two aspects are both necessary and supplement each other.

Thus, although scientific and technical training is essential for activities like engineering, humanistic training helps to improve the social attitude of professionals. Improved imagination, mental versatility, and the ability to see analogies between apparently different fields may offer new inspirations even for the development of a technical specialty.

On the other hand, scientific and technical training are equally necessary to the humanities, as they offer a method that may be applied in a constructive way as cultural assets for the benefit of mankind. Scientific training offers us a safe method for coping with the concrete, specific problems that we have to face every day; sometimes an exclusively humanistic approach tries to solve everything in terms of principles and fails to tackle problems in a practical way.

We must not expect from science or technology more than these disciplines can give us. They can provide us with effective methods for the control of the forces of nature. However, they cannot provide us with a philosophy of life, nor the wisdom with which better to utilize those forces. Obviously, one should not underestimate technical competence and scientific knowledge which must go deeper and enrich our way of looking at the world. Although science and technology are ethically neutral, they can be used either to construct or to destroy. However, they cannot tell us which of these alternatives must be chosen. Therefore, the continuous progress of science and technology emphasizes

---

*Material presented by the author has been subject to extensive editorial revision.

the need for moral intelligence in man and for humanitarian values in society. Thus what is needed in education is a balanced approach that emphasizes the humanistic disciplines and at the same time develops proficiency in the sciences. We must teach the students not only science but also history, philosophy, ethics, and the arts. We must create a closer connection between the scientist, the philosopher, the historian, the artist, the theologian, and so on.

The question is how may we achieve these goals. We are well aware that our curricula are overburdened. We should not gain much by having a greater number of classes. I am of the opinion that the basic study of all these humanistic disciplines, science, history, languages, ethics, and art should be carried out mainly during secondary education, which should provide a solid training for a more generalized culture. On the university level, the broader aspects should be considered from the philosophical, historical, and social viewpoints, as well as artistic and theological viewpoints. Here we must not forget that the true, sensitive, and perceptive attitude, the humanistic feeling which is necessary, originates from the interior growth of ideas.

The importance of an inspiring atmosphere is well known. Thus instead of having separate buildings in which only professional training is administered, as is the case in South America, we should try to give the students the true leisure that a university campus can give, where they can have their professional training and at the same time their extracurricular activities. There they would have their own cultural community and the opportunity of continuous contact with teachers.

In conclusion, I should like to quote from Plato the following words said by him about 2500 years ago:

"It is not the dispensing of knowledge, even if it would include all the sciences, which creates welfare and happiness; but rather a particular sector of knowledge: that of good and of evil. We might argue that medicine can give us health, the art of the shoemaker gives us shoes, the art of the weaver gives us clothes, the science of navigation can save lives on the sea, and strategy can win battles for us. But, if we exclude the knowledge of good and evil we will find that the very use and the excellency of these sciences will have failed us."

3

The Contribution
of Physics to
Liberal Education

THE AIMS OF ELEMENTARY- AND

SECONDARY-SCHOOL SCIENCE*

Professor A. M. J. F. Michels (The Netherlands)

In this paper we are concerned with the goals in education which parents and society have entrusted to our elementary and secondary schools. What is the target for our education of these children, and what part can science play?

The students considered here are children from kindergarten to the age of sixteen, although we recognize that many children terminate school education earlier because of their lack of innate ability, a lack of educational facilities, or a lack of stimulation in their home environments.

At the end of the school period the foundation must have been laid so that, when grown up, the young man or woman can live a life where he or she can feel satisfied and can be a useful member of society. It is often explained as emphasizing that society determines the child's future, while in our opinion in a democratic country society is built by its members; they form society and not the other way around. An honest, prosperous, homogeneous society is based on its citizens, whose inborn good capabilities, mental, intellectual and moral, are well developed by a balanced education. For this, general education must lay the foundation.

A man is a rational being. He has the power to think, the desire to create. Inborn is the love for the aesthetic appreciation of what is beautiful, be it in creation or in the products of the human mind. If these gifts, received at birth, are neglected, not developed by education, his road towards complete happiness is certainly endangered if not blocked.

From these considerations we draw the conclusion that, whatever the knowledge of facts or the technical skill required by a future

---

*Material presented by the author has been subject to extensive editorial revision.

profession, from a personal point of view the development of the intellect by broad general education is of primary importance. This claim does not end at the age of sixteen or when the child leaves the school earlier to enter into a trade.

It is gratifying to realize that society also benefits from an education along these lines. Nowadays, society develops so rapidly that a major part of what is learned at school is already outmoded by the time the adult has become an active member. What remains of value is the ability to observe, to think logically, to draw conclusions, and to make decisions. In our rapidly changing world, decisions must be based on an understanding over a broad field, an appreciation of what a whole line of thought can contribute, how other people feel and think, and an evaluation of aesthetic and general cultural values.

Here a school is confronted with another obligation. Besides the classical subjects of the curriculum — reading, writing, and arithmetic — the school must provide the means to make responsible choices. The pupil must have an inkling of what is going on in the world, as far as his intellectual capabilities allow him to grasp it. He must be able to form an opinion of what can interest him in his adult life, not only for a job but also for his hobbies, and especially hobbies with an intellectual background in which he can take refuge if the humdrum of daily life threatens to overwhelm him. He must have learned to enjoy art, mental and physical exercise, to be unafraid of silence or of being alone with his own thoughts. He must learn that in the world around him many things are exciting. One of the biggest crimes a teacher can commit is to make his subject dull.

We come now to the cultural aims of education and we are immediately confronted with a serious difficulty. If we follow for a moment the line adopted by C. P. Snow, who cannot be denied the merit of drawing the attention of the world to the existence of controversy, there exist two cultures, one human, the other scientific. Both can be regarded as achievements of the human mind. Whether other results of human mental power and intelligence, economics, sociology, history, and commerce may be classified amongst one of the two mentioned, or considered separately, need not be discussed. For the sake of simplicity we shall restrict ourselves to the humanities and the sciences.

What have scientists and humanists in common which can help toward mutual understanding?

Let us identify some of the activities of a humanist and take as an example the study and use of a language. Anyone who wants to use a language must know and memorize some facts, the meaning of the words, the laws of grammar, and the exceptions to these laws. This is the first phase. The second phase is the use of the language for communication, for speaking, writing, and reading. This use of the language is an application for practical purposes. It provides the opportunity for contact between nations, betweeen governments and people, between industries and consumers; it has enormous economic value, at least as great as that of new inventions derived from our increasing knowledge of nature. For centuries we have been so used to this economic tool that we do not realize what we should miss if languages did not exist. Through daily use for tens of centuries the glamour, though merited, had faded away.

However, the humanist's final aim is literature. Here, in the third phase, is where real culture is involved. The language itself withdraws to the background and becomes only an indispensable tool to record and transmit the products of the artist's mind. A similar analysis could be given of all branches of art.

What parallels between science and the humanities can be drawn, recognizing, of course, that no analogy will be perfect or complete?

The scientist uses a knowledge of facts and rules that he calls laws. These laws seem to be more binding than the rules of grammar. The number of exception is smaller, and if he finds any, he is inclined to look for a reason. The reason for this attitude is the result of experience through the ages. There was a time when most phenomena were not understood and were blamed on evil spirits or credited to benevolent ones. As the scientist discovered fundamental laws, the belief in spirits tended to disappear, and he became convinced that every phenomenon of nature was determined by absolute rules. Rationalism and determinism thus emerged in the middle of the last century. This conviction still holds for macroscopic or submacroscopic phenomena, but has been shaken for the world of the submicroscopic and elementary particles by the uncertainty principle of Heisenberg and by the development of wave mechanics. In these fields probability has replaced the old strict laws of determination.

Thus in a first phase the student must learn the fundamental facts and laws and some experimental techniques in order to observe and analyze phenomena of nature.

The second phase is the use of science for practical purposes, technology with all its impact on present-day life.

Now comes the third phase — that of creative work in science and appreciation of its meaning. This is where it becomes a cultural activity.

How does the scientist work? He takes his observations, which may arise naturally or as the result of special ad hoc experiments. In his experiments he puts questions to nature and hopes to receive an answer; the formulation of the questions put to nature, or, in other words, the conception of the experiment is primarily a result of the intuition of the scientist and of the knowledge he already has.

What does the scientist, a physicist perhaps, do with his observations? He tries to present them in such a way as to see whether a common aspect can be discovered among the answers. If by varying the circumstances he always obtains the same effect, he calls it a law. In the formulation of this law, however, intuition and abstraction are in general involved; at the same time a model is introduced. In the formulation of Newton's law (force is equal to mass times acceleration) the concepts of force, mass, and acceleration were established; absolute time and the existence of a system of coordinates were accepted. This model proved to be of high value in the early development of physical science. That it is only an approximation became apparent only much later, as any scientist knows.

In the formulation of a law causality must be accepted. For any action there must be cause; the cause of the acceleration is the force.

The model created by the scientist also serves other purposes. Besides being used as a sort of language to express the results of his observation as a coherent picture, it is also used to make predictions. If in a new series of observations some details seem to fit an already existing model, the scientist examines whether other facets of his model could be fitted to the new phenomenon under study. Here his intuition plays an important role. Model and intuition lead him to design new experiments. If the prediction of the model does not agree with the new data, the model is changed or a new model is created. Gradually the new model loses contact with reality, it becomes a conception of the scientist's mind, in a way not unlike the path the modern artist follows. In practice a model is in the form of a mathematical description. The pragmatic value of mathematics, both as language and as model, comes to life. It is appropriate to stress here that over and above this pragmatic value, mathematics is an art in itself which ranks at least equal with the other sciences.

As long as the sciences deal with macroscopic phenomena, the difficulties are not excessive. The picture becomes much more complicated when studying the submicroscopic world. We know nowadays that our instrumentation cannot give us decisive information about atoms, electrons, and elementary particles. We cannot make a picture of these particles. Accustomed to think of macroscopic phenomena, our imagaination falls short, and we know it. Still, the value of the old model is very real.

Meanwhile, the new language, the language of mathematics, is further developed. Where relations in nature are always found to have both qualitative and quantitative aspects, it is obvious that mathematics plays an important and nowadays even a predominant role in the description. More and more, mathematical expression is taking over the role of the model.

We now return to the similarities between the humanistic and scientific activities. The parallel of the corresponding three phases in the two fields is clear from what has been said. An even greater similarity becomes obvious when we consider the act of creation, of which the human intellect alone is capable. The author of a fine essay or book, an artist who produces a painting or a statue, a theoretical physicist who discovers a new theory, an experimentalist who devises a new method to reveal one of nature's yet unknown secrets, and even the technologist who invents a new engine, all perform acts of creation and all experience the special satisfaction of creativity.

Scientists are constantly blamed for discoveries that can lead to disaster as well as to human benefit. This is also true of the humanist's activities, particularly those related to the use of language. The arts of speech and writing have been misused for ages, frequently by demagogues for purposes quite inhuman. We have only to refer to the origin of the devastation and murders during Hitler's regime. Also, pornography would be impossible if writing and painting did not exist. Whether the moral disaster following the pragmatic misuse of the language is less than the materialistic attitude in the wake of technology or whether the invitation to destruction by devilish propaganda is less serious than devastation by armaments are questions we may never resolve. The point is that misuse of the products of the human mind can be found everywhere.

The true fundamental scientist is a humanist too. But he must live up to it. As a humanist he must not disdain other branches of humanism, a disdain that hurts his colleagues of the classical arts. It is proper that in the presentation of his work he should demonstrate the same logical approach, and in the language of the spoken word the same style and exactness of expression which he loves in his mathematical formulation. The other group, however, must appreciate that there is more in science than its material application, and must be aware that their own culture has produced pragmatic values at least as important as those of the sciences.

If we return now to the point from which we started and consider C. P. Snow's statement that there are two cultures, that of the humanist and that of sciences and technology, two points become clear. The first is that Snow tried to compare noncomparable systems, two systems of different dimensions. For the humanities he placed the emphasis on the third facet, which our analysis described as the real essence of art. For science he stopped at the second facet, the pragmatic application. It is interesting to note that when he looks around for a medicine to cure the situation he again takes refuge in a pragmatic solution. Despite the admiration we have for his suggestion, "Let the countries who still suffer under poverty share in the wealth produced by our technical effort," we feel that in this suggestion, although certainly to be followed up, he missed the most essential point, which is that education in science may be developed in a way that leads to an appreciation of its cultural value as well as of the pragmatic applications, in exactly the same way as humanistic education. Above all, it should cultivate the interest and curiosity of the child and lay the foundations so that they are never lost.

When we try to find what is good and bad in our present society and search our consciences to discover whether our education system is partially responsible for what we find, we must ask ourselves a number of questions. Do our schools merely train children for society or do they develop inborn abilities and prepare children for adult life? Does our education train our pupils to be logical, rational beings, whose intellectual power must be given first attention, or do we drown them with barren facts and laws? Do we approach our pupils with the marvels of nature and acquaint them with modern interpretations as far as they can grasp them, or do we stuff them with an old curriculum? Do we appeal to their understanding or to their memory? Do we train them to solve arithmetic problems without much intellectual value? Do we forget that up to the age of sixteen, a foundation must be laid on which to decide in general terms the direction of adult life? Do we show them that science and arts are exciting?

As we balance the education of our children, we must emphasize the quality of logical, rational thinking, an attribute of the cultured man more important and more durable than any collection of facts and information which he will live to forget.

In conclusion, we quote from a recent lecture by J. H. London, president and director of the Shell Oil Company:

"In recent years two paradoxical anxieties about our education have found expression. The first is that we do not educate enough technicians to provide for the maintenance of culture. The second, too much emphasis is given to the sciences and technology at the cost of the humanities, to the detriment of our culture. It is my conviction that with a small amount of logical reasoning and a minimum of dogmatic statement the contradictions will disappear. Science and the humanities are not incompatible. In our industry, in particular, we need technically educated people aware of the humanities and also those educated in the classics, who have an understanding of the sciences.

"In the complicated structure of the lines of communication and the interplay of the branches in our industry we receive the most valuable contribution to good management from those scarce individuals who are at home in both worlds, the scientific and the humanistic."

THE GOALS FOR SCIENCE TEACHING*

Professor Gerald Holton (U.S.A.)

In his speculative essay, "The Rule of Phase Applied to History," dated January 1, 1909, the American historian, Henry Adams, came to a remarkable conclusion: "... the future of Thought, and therefore of History, lies in the hands of the physicists, and...the future historian must seek his education in the world of mathematical physics. A new generation must be brought up to think by new methods, and if our historical departments in the Universities cannot enter this next Phase, the physical departments will have to assume the task alone."

In arriving at this startling view, Henry Adams explains that he was guided by a desire to transfer to the study of history some of the conceptions developed in 1876-1878 by Willard Gibbs, Professor of Mathematical Physics at Yale, in his famous paper "Equilibrium of Heterogeneous Substances," and by other physicists and chemists who followed him. Just as the physicist Storey saw in the Phase Rule a means for putting into hierarchical order the sequence of "phases" consisting of solid, fluid, gas, electricity, ether, and space, so did Adams believe that thought, too, in time had passed through different phases. Acknowledging himself to be a follower of Turgot, Littré, John Stuart Mill, Comte, and others who had also held that history obeys quasi-physical laws, Adams found confirmation for the correctness of his essentially prophetic and apocalyptic view of history, in the apparently increasing rate of change of historic processes, and he thought they were analogous

---

*Material presented by the author has been subject to extensive editorial revision.

to the increasing motion of objects that are attracting one another by an inverse square force, or to increases in the rate of other similar physical effects.

I have cited this not because physicists would believe that history obeys laws closely analogous to those of physics — indeed, a meeting of physicists, such as this, would be the last place to find sympathy with the modern-day physiocrats. Rather I have spoken about Henry Adams because, as usual, he had brilliant insights in this essay, primarily on two counts. He said that the ideas emerging from physics would continue to be, as they had again and again been since the seventeenth century, a central part of modern culture; and, second, he saw that science is both a major mechanism for change in culture as well as a way of understanding the change better.

Today, we should like to believe that these insights are generally shared by all men who have thought about the matter. Thus the chairman of the National Commission on College Physics in the U.S.A., Professor Walter Michels, has recently said:

"The recognition of the role of physical science as a chief determinant of our culture, and the recognition of physics as a discipline having many of the educational values associated with the classics in the nineteenth century are contributing to the current trend.... The community of academic physicists will be hard pressed to meet all the demands made upon it, yet it would be shirking its responsibility if it failed to do so."[1]

And in a completely analogous way, Professor A. S. Akhmatov of the U.S.S.R. delegation to the 1960 UNESCO Paris Conference on International Education in Physics stressed the basic place of physics for the general cultural requirements of educated men and women:

"It is clear at present to everybody that it is precisely the progress of physics that determines the possibilities of development in a very wide range of sciences, from cosmology to biology and medicine. It is physics that determines to a large extent the foundations of our outlook as well as the possibilities and limits of our practical activities. One cannot be called a specialist or, for that matter, an educated person unless one is familiar with a certain range of ideas and facts in the sphere of physics".[2]

Thus, despite differences in local custom or ideology, we are likely to judge and express the matter the same way; but this is by no means generally agreed to outside these walls. There are two main groups — not to mention the philistines who have always viewed any intellectual activity with suspicion — that would deny the proposition that physics has a central position in our culture. One represents, as it were, the

---

[1]"Progress Report of the Commission on College Physics," Am. J. Physics, 30, 17 (October, 1962).
[2]Quoted in S. C. Brown and N. Clarke (Eds.) International Education in Physics, Proceedings of the International Conference on Physics Education, UNESCO HOUSE, Paris, July 18th–August 4th, 1960, The Technology Press and John Wiley & Sons, New York, 1960, p. 129.

reaction from the right, from the side of misguided traditionalism. They would say with Matthew Arnold[3] that "culture is, or ought to be, the study and pursuit of perfection," and would then define the properties of perfection — for example, beauty and intelligence — in such a way that most scientific work stands exposed and condemned as soulless hack-work and the manipulation of trade-school mechanics. Or with T. S. Eliot,[4] they would say that culture and religion are "different aspects of the same things," and then, instead of noticing that science is also an "aspect of the same things," they would then define culture and religion in such a way that science, when it is mentioned at all, becomes identifiable with idolatry.[5]

A revealing quotation from an editorial in the English Sunday Telegraph of 11 March 1962, cited with shock by my colleague John H. Van Vleck in his fine essay The So-Called Age of Science,[6] provides evidence for this point as clearly as we can wish:

"A free and prosperous society depends on the activities of three distinct classes — a political elite trained by the study of the humanities to take broad and enlightened views about ends and means, a technical elite, willing to exercise its skill in obedience to the community's will, and a proletariat with enough mechanical intelligence to respond to managerial direction.

"The first of these conditions has not yet been entirely removed by the renunciation of the classics, the second will be secure so long as scientists are not encouraged to dabble in the humanities to develop a wish to govern society, and only the third ... is strikingly absent...."

The other, or left-wing, opposition comes from the ranks of science itself. A relatively small but influential proportion of scientists would say that there simply is no meaning in the proposition that physics, for example, is more than what they and their best associates are actually doing at the blackboard or in the laboratory here and now. Unlike the first group of opponents who object that science has little place in culture, the second objects that culture has little validity compared to science. Only politeness prevents them from dismissing with vocal impatience the suggestion that there are valid and important links between what happens in the laboratory now and what happens, has happened, and will happen elsewhere — on the stage, in the sculptor's

---

[3]Culture and Anarchy, Macmillan, New York, 1883.
[4]Notes towards the Definition of Culture (1st Amer. ed), Harcourt Brace, New York, 1949.
[5]That such feelings are not confined to intellectuals, but are more widely shared, has been shown in a number of studies, for example, the extensive research by Donald D. Dowd and David C. Beardslee, "College Student Images of a Selected Group of Professions and Occupations," Wesleyan University (Middletown, Conn., April 1960) (lithoprinted); some of the main conclusions were summarized by these authors in "The College-Student Image of the Scientist," Science, 133, 997-1001 (1961).
[6]Cherwell-Simon Memorial Lectures, Oliver and Boyd, Edinburgh, 1962.

studio, in the courtroom, in the study of a philosopher or an economist, and even in the nursery where a child is asking his mother for help in making sense out of the world around him. And the major reason why some of these scientists can neglect the complex, tenuous, long-range links that attach to their science is that they are so successful doing what they are doing. The short-range forces, which they master, completely saturate their capability for forming and perceiving long-range connections.

One is reminded of the diagnosis C. P. Snow offered in his provocative book Science and Government for the reason why some scientists so single-mindedly stuck to a narrow decision or were satisfied with narrow range of investigations. It was their success in one particular field or with the operation of one particular apparatus. Snow dubs these men "gadgeteers." But at this Conference on Physics in General Education, I shall be safe in assuming that one does not have to defend further the proposition of centrality of science in culture, either from the radical scientism of the left or the culture snobbism of the right. On this middle ground, we shall thus posit that the much-discussed cleavages of knowledge are all too often the unhappy results of erroneous definition. The controversy between T. S. Eliot and his critics, or between Snow and Leavis more recently, serves to remind us how necessary it is for each age to rethink what "culture" is in each of its multiple senses, what makes the culture of a people cohere, and what forces and mechanisms are at work to change it. In this light the important topic is not to what extent science is separated from other activities, but rather how we may define and transmit culture in such a way that the sciences are seen to be valid components of our culture. We therefore must here adopt, as one of our main tasks, the actual design of educational curricula stressing the coherence of physics and the other components of intellectual life.[7]*

Threats to Coherence

Before any specific proposals can succeed, we should be able to deal with the major threats which every coherence-seeking program of science instruction will face. Two of these threats should now be discussed briefly.

---

[7]A discussion of detailed goals for such courses has recently been given by the author in "Science for nonscientists: Criteria for college programs," Teachers College Record, 64, 497-509 (March, 1963).
*The design of an actual curriculum useful at the university level was discussed in a paper submitted by Professors A. Harasima and D. C. Worth (Japan) under the title "A New Natural Science General Education Program for Physics and Other Science Majors at International Christian University (Tokyo)." The course discussed was for fourth-year university students, where the opportunity was given for studying the basic nature, philosophy, and methodology of that area in which the students are majoring, and to study the position of that area in general human culture and in the activity of mankind as a whole.

The first is the rapidity with which the simplest terminology of the contemporary sciences is being removed from the natural language of the beginner. This makes obvious difficulties for the new learner — not to speak of the difficulty this student will have ten or twenty years after he has passed through our classroom and, at the height of his career and ability, faces a science that is by then concerned with entirely changed problems phrased in an entirely changed vocabulary. As we here all know very well, for his teachers this is also a major and continuing problem. But on an even more fundamental level, there is the additional and obverse problem of the effect of the increasing vocabulary gap, not merely on the new learner, but on the language of science itself.

This is indicated by an obvious difference between even a most difficult and sophisticated natural language, such as Sanskrit, and even a relatively simple contemporary science, such as introductory quantum physics; the difference is clearly that the difficult natural language can be, after all, learned by any small child, whereas the simpler modern sciences turn out to be all but inaccessible to many of our most intelligent students. The point is of course that natural languages have developed in a very different way from scientific language. As Margaret Mead perceptively noted in an essay from which I wish to quote at some length,[8] "it has been characteristic of all earlier forms of cultural transmission that new intellectual acquisitions — such as script, mathematical calculation, prosody — have been taught in face-to-face situations by adults who knew, to children and adolescents who did not yet know." Miss Mead finds that while in the past even contemporary scientific language used to be so communicated, this situation has begun to change radically with the rise and acceleration of the sciences in recent times. "As the several sciences have begun to grow at an unprecedented rate, the distance from one major advance to the next is often reduced from fifty to five years. And now, instead of teaching a widely selected, intelligent student audience, more and more the young scientists are communicating to each other, horizontally, in highly specialized languages, material so new that publication is not rapid enough to encompass it.... We are, in fact, in danger of developing — as other civilizations before us have developed — special esoteric groups who can communicate only with each other and who can accept as neophytes and apprentices only those individuals whose intellectual abilities, temperamental bents, and motivations are like their own. A schismogenic process is under way that is self-perpetuating and self-aggravating....

"All of us who cherish the change in pace made possible by this new kind of horizontal, face-to-face, multimodel transmission, which works even across national boundaries, inevitably will guard jealously any attempt that would seem to slow up this intoxicating process. But now we must find new educational and communication devices that will not sacrifice this new high level of specialized communication and yet will

---

[8]Margaret Mead, "Closing the Gap Between The Scientists and the Others", Daedalus, 88, 139-146 (Winter, 1959).

protect our society and all the intellectual disciplines within it from the schismatic effects of too great a separation of thought patterns, language, and interest between the specialized practitioners of a scientific or humane discipline and those who are laymen in each particular field."

This is, I believe, an important warning, and a challenge that may be directed to us here more properly perhaps than to most other gatherings of scientists. It gives us an additional reason for insisting that our young students should receive early and full opportunity to learn the concepts and theories of modern science, and so to be brought to a state where the vocabulary and grammar of modern science, including some of the techniques of calculation, will no longer themselves be the main obstacle to an understanding of the proud achievements of our time. And Miss Mead concludes that such a program will also modify the threat that science will escape entirely from the area of natural discourse:

"The process of vertical communication of results arrived at by horizontal, face-to-face adult learning will alter the vocabulary and syntax of the <u>communicators</u>. Thus they will be the more able to transmit what they know, and they themselves will keep in closer touch with the other specialties of our highly specialized societies. Any language taught only by adults to adults — or to children as if they were adults — becomes in certain respects 'dead.' It fails to enlist recruits, it may lose its productivity, and it serves in the end primarily to separate those who know it from those who do not. In contrast, any language that is taught to all children attains a multimodel comprehensiveness that makes it a suitable vehicle for the thought of not only the highly intelligent but also the moderately intelligent and the deviantly endowed person. By insisting that all children, not only those children who, by joining the ranks of a discipline, will accentuate its highly specialized style, should be taught recent advances in a particular discipline, we can set up an automatic corrective system for the dangerous intellectual divergencies of vocabulary and knowledge within our society...."

The thought of our most creative scientists exposing themselves to the rigor and danger of an elementary classroom is, happily, no longer implausible. More and more this hope is being realized, and particularly in physics teaching. Without doubt, the benefits to students and science itself are going to be even larger than those we have just begun to harvest from this movement.

We turn to the second threat to the achievement of coherence, and here we may well wish that Miss Mead's warning had been heard and heeded long ago. For the failure of "the others" to understand what the creative scientist now knows and does has had, I fear, some debilitating and even tragic consequences — particularly to the effectiveness and morale of our most valuable intellectuals outside science. For they are caught between their irrepressible desire really to understand this universe, and, on the other hand, their clearly recognized inability to make any sense out of the simplest vocabulary of modern science. I have stressed elsewhere[9] the chilling realization that our intellectuals,

---

[9] In the essay "Modern Science and the Intellectual Tradition", <u>Science</u> <u>131</u>, 1187-1193 (April 22, 1960).

for the first time in history, are losing their hold of understanding the world:

"The nonscientist realizes that the old common-sense foundations of thought about the world of nature have become obsolete during the last two generations. The ground is trembling under his feet; the simple interpretations of solidity, permanence, and reality have been washed away, and he is plunged into the nightmarish ocean of four-dimensional continua, probability amplitudes, indeterminacies, and so forth. He knows only two things about the basic conceptions of modern science: that he does not understand them, and that he is now so far separated from them that he will never find out what they mean."

To take a concrete case, consider for example the recent, widely read book The Sleepwalkers,[10] by Arthur Koestler. In it Koestler tries to trace the rise of modern physics, and, with it, of modern philosophical thought, stemming from the work of Kepler, Galileo, Newton, and some of their contemporaries. This is indeed still a useful task to set oneself. Koestler has worked with devotion on his material. And, most important, he is of course the intelligent layman par excellence whom any scientist would be pleased and proud to have as his pupil in this evidently earnest search for an understanding of modern science.

And yet, something terrifying happened as Koestler came to the end of his book. He had still been able to see meaning and order in the physics of the seventeenth century. When he turned in the Epilogue to modern physics, all sense of understanding and coherence disappeared, and the incomprehensible modern conceptions seemed to rise around him on every side as threats to his sanity. As he summarizes his work, he finds that to a large degree "the story outlined in this book will be recognized as a story of the splitting-off, and subsequent isolated development, of various branches of knowledge and endeavor — sky-geometry, terrestrial physics, Platonic and scholastic theology — each leading to rigid orthodoxies, one-sided specializations, collective obsessions, whose mutual incompatibility was reflected in the symptoms of double-think and 'controlled schizophrenia.' "

I believe it is important to consider this case as sympathetically as we can — to listen to the anguish of an intelligent man who has discovered that he cannot cope with the modern conceptions of physical reality. For what he is saying to us is what most people would say — if they were eloquent enough and interested enough in knowledge to be deeply disturbed by a state of unchangeable ignorance.

"Each of the 'ultimate' and 'irreducible' primary qualities of the world of physics proved in its turn to be an illusion. The hard atoms of matter went up in fireworks; the concepts of substance, force, of effects determined by causes, and ultimately the very framework of space and time turned out to be as illusory as the 'tastes, odours and colours' which Galileo had treated so contemptuously. Each advance in physical theory, with its rich technological harvest, was bought by a loss in intelligibility...."

---

[10]The Sleepwalkers, Macmillan, New York, 1959.

"Compared to the modern physicist's picture of the world, the Ptolemaic universe of epicycles and crystal spheres was a model of sanity. The chair on which I sit seems a hard fact, but I know that I sit on a nearly perfect vacuum.... A room with a few specks of dust floating in the air is overcrowded compared to the emptiness which I call a chair and on which my fundaments rest....

"The list of these paradoxa could be continued indefinitely; in fact the new quantum-mechanics consist of nothing but paradoxa, for it has become an accepted truism among physicists that the sub-atomic structure of any object, including the chair I sit on, cannot be fitted into a framework of space and time. Words like 'substance' or 'matter' have become void of meaning, or invested with simultaneous contradictory meanings...

"These waves, then, on which I sit, coming out of nothing, travelling through a non-medium in multi-dimensional non-space, are the ultimate answer modern physics has to offer to man's question after the nature of reality."

And at the very end of the book, in its last, furiously splashing paragraph, I cannot but hear the cry of a drowning man, a cry for help that cannot leave one unconcerned if one believes that physics can and must be shown to play a valid, creative part within our culture:

"The muddle of inspiration and delusion, of visionary insight and dogmatic blindness, of millennial obsessions and disciplined doublethink, which this narrative has tried to retrace, may serve as a cautionary tale against the hubris of science — or rather of the philosophical outlook based on it. The dials on our laboratory panels are turning into another version of the shadows in the cave. Our hypnotic enslavement to the numerical aspects of reality has dulled our perception of nonquantitative moral values; the resultant end-justifies-the-means ethics may be a major factor in our undoing. Conversely, the example of Plato's obsession with perfect spheres, of Aristotle's arrow propelled by the surrounding air, the forty-eight epicycles of Copernicus and his moral cowardice, Tycho's mania of grandeur, Kepler's sun-spokes, Galileo's confidence tricks, and Descartes' pituitary soul, may have some sobering effect on the worshippers of the new Baal, lording it over the moral vacuum with his electronic brain."

This is the end of his road that started with great hope and promise. What shall we do about it?

Promotion of Coherence

There are several related areas of study that would alleviate the difficulties we have discussed and promote the sense of coherence which we seek to cultivate.

The first, obviously, is the study of modern physics itself. Miss Mead's and Arthur Koestler's arguments both speak first of all for substantial attention to modern physics in the school and college curriculum.

I have little use for the attitude that one hears occasionally, to the effect that quantum physics and relativity should precede or even take the place of classical physics in introductory courses. I believe that this is pedagogically unsound for most students, that we must help their intuitions to bridge the distance from their natural Aristotelian bent to the strange new conceptions by means of classical (Galilean-Newtonian-Maxwellian) ideas. The growth of imagination recapitulates the growth of a field to a large extent.

But whether we all agree on the order of presentation or not, the main point should be beyond dispute here. The college student should, as part of his general education, arrive at an understanding of the main concepts and theories of modern physics. This includes certainly an introduction to quantum theory and relativity theory. Such a program necessitates careful consideration and use of the whole curriculum, from grammar school to college, in order to provide enough sophistication and tools, both in mathematics and in physics, to ensure that the minimum end-point achievement in physics be not superficial. Nevertheless, this effort can, I feel, no longer be shirked. In our world, the essential notions of quantum physics and relativity are no longer optional for anyone who wishes to regard himself as generally educated, any more than are the essential notions of certain specifiable other fields.[11] It is also important to design impedance-matching devices in our curriculum by which a student from such a serious general education course can continue in more advanced physics courses if he wishes. Regardless of his field of final concentration or specialization, it should be possible for the talented and interested student to receive physics instruction at the highest level of which he is capable. The administrative problems for doing this are large at most colleges in the U.S.A., and worse at colleges following the European model. But this merely shows from another angle what all of us know already; that the ever-waiting administrative difficulty is the last thing to consider and the first thing to stamp out ruthlessly if any sound idea is to come to anything beyond mere talk.

Difficult though a program of serious physics instruction for all college students is to carry out, it represents in my view only half the necessary job. Even if we succeed in presenting fairly sophisticated physics to students who will not go on in the sciences, but have done nothing more, we shall have failed. For these students by and large do not come to us as we ourselves came to the study of physics. We, who are all trained as physicists, demanded no more than good physics from our introductory courses. They, however, who do not intend to become us (sometimes a difficult point to remember!), want to see also what place physics has in the total reality, in the context of all intellectual endeavors; and unless we help them, nobody will, and they will know that they came to our shop erroneously. We should remember a story C. N. Yang told not long ago to express the disenchantment of some

---

[11]E.g., in a biological science, an analytical social science, a historically oriented social study, a literary humanistic field, and a creative arts field — to indicate by only one phrase each of the other main subjects in which a sound general-education experience seems to be now essential for any college student.

physicists with the mathematician who is not interested in the realities of our concerns. "There is a story circulating among us," Professor Yang said, "describing the feelings of a physicist when he consults a mathematician. A man carried a large bundle of dirty clothes and searched for a laundry without success for a long time. He was greatly relieved when he finally found a shop displaying a sign 'Laundry done here' in the window. He went in and dumped the bundle on the counter. The man behind the counter said,

" 'What's this?'

" 'I want to have these laundered.'

" 'We don't do laundry here.'

" 'But you have a sign in the window advertising that you do laundry.'

" 'Oh! That! We only make signs.' "

We, too, are all too often only making signs, if we teach our non-physics students as if they are going to be scientists.

To stress the wider context of science, some college courses have attempted to go away from science itself and have instead stressed the study of joint areas involving science as one partner. I distinguish two kinds: " 'something' of science" courses, and " 'something' and science" courses.

The first of these types yields courses in history of science, sociology of science, philosophy of science, and so forth. They are characterized by using the activity and accomplishments of science as the raw materials, upon which to operate with the scholarly tools of history, or sociology, or philosophy. This is of course a worthy and useful procedure, but only if the raw material, namely science, has been fully understood. Following the study of science (for example, in the syllabus of a science concentrator in his last year or two, or as a professional study in its own right, undergirded by thorough science courses) it can be an interesting field. But as an alternative to the study of science, such courses seem to me dubious if not misleading, and particularly so for the introductory student.[12]

The second of the joint-area studies, the " 'something' and science" study, can be on history and science (as in the case of Henry Adams' essay), or psychology and science (as in the studies by A. Roe and B. Eiduson), or science and public policy (A. H. Dupree, or Donald Price, or J. S. Dupré and S. A. Lakoff), or science and art (E. Panofsky or J. Ackerman). These do not use the achievement of science as raw material to as large a degree as the previous study area, but they are likely to develop the link between the two members of the couple without subjugating one to the other.

These studies, too, cannot be regarded as replacements for sound science study, but they can also be of great help in the coherent presentation of physics in general education. Indeed, the specific suggestion

---

[12]This is not said to condemn another use of the history or philosophy or sociology of science, namely to aid in the presentation of the material of science itself in a science course. On the contrary, I have long both preached and practiced the doctrine that these aids are of great interest and use to the student.

I have to make is based on the idea of introducing into the conventional physics course, which we all know well how to give, a multiplicity of references, illustrations, short discussions, longer reading and essay assignments, all based on various joint-area studies. Thereby we can build into our course the spirit as well as the substance of the coherence — making ties by which our students can clearly see from the beginning that physical science does not stand as an isolated and forbidding area, having no relationship or analogies with other fields. On the contrary, in this way they should be helped to see clearly what most physicists themselves are apt to take for granted without making the effort to verbalize it, and what is so easily lost in the haste of most "straight" physics courses: that any part of present-day physics has important connections with other achievements of human beings, with sciences other than physics, and with studies and activities other than science.

For the purpose of the discussion of such a course, it is useful first to think of the physics course that is traditionally given at many colleges, in terms of a convenient model. The coverage to be attempted in a physics course is represented in Figure 1 on a three-dimensional "map." On the x axis lie the academic fields that make up the total academic educational experience of the student, arbitrarily arranged from the most quantitative (mathematics) to the least (humanities). The y axis may represent time from, say, 1600 to today, which, for introductory physics courses, corresponds very roughly to the development of the usual subdivisions (mechanics to nuclear physics). The z axis represents depths of penetration attempted in the course on a given topic. On this map the physics course usually given today may be represented roughly by a set of closely spaced and sometimes overlapping pyramids, based in a more or less narrow and well-defined strip in the xy plane.

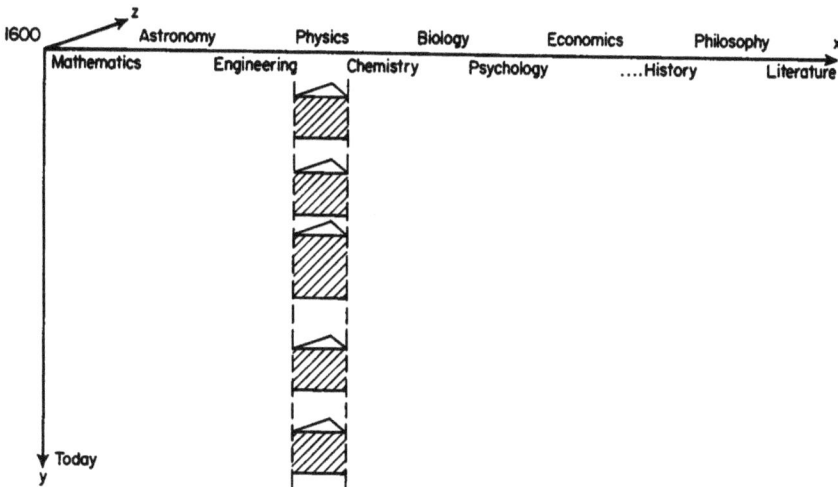

Figure 1. Traditional presentation of introductory physics

This narrow-focus scheme may be defended in course directed to preprofessional physics concentrators, but not for "the others." For them a variety of courses have been invented and are being offered in many of the colleges. I have provided elsewhere a summary of the types of orientation courses (as opposed to preprofessional training courses) that have been found more or less successful in the U.S.A.[13]

The Connective Physics Course in General Education Programs

When a physics student goes on to higher level science courses and finally to research, he eventually sees that the field he is studying does hang together. For as his study of physics and related sciences penetrates further along the axis of depth in Figure 1, he unavoidably finds that the separate "pyramids" representing mechanics, optics, and so on, in Figure 1 are ultimately joined together along the depth dimension in a single interrelated corpus of knowledge that also stretches far to the right and left of the narrow strip under "physics." Although as an undergraduate physics student he will have taken mathematics, engineering, and chemistry courses in different departments and buildings, as soon as he tries to do any significant piece of research he discovers that the separation between neighboring sciences was a pedagogic, administrative convenience, which from the point of view of living science is just as artificial as the separation between the "pyramids" in Figure 1.

Thus we find of course that an experimental research project on, say, the dependence of molecular relaxation effects on pressure involves sooner or later material from every separate block in a physics course, but also brings in mathematical methods, metallurgy, electronic engineering, chemical thermodynamics (not to speak of the areas of politics and psychology which anyone must cope with if one wishes to obtain and correctly administer research grants in a busy department). No one who has engaged in actual scientific work can fail to have seen the intimate connection between physics and advances in other sciences and engineering or between the advance made in pure physics and its social and other practical consequences. Indeed, "pure" physics is an invention that exists only in the old-fashioned classroom. As soon as a real problem in physics, or any other field, is grasped, it appears that there hang from it connections to a number of expected and unexpected problems in fields that, by habit, we make our students think of as "belonging" to other professions.

Now while our preprofessional students discover by later experience the existence of these connections sooner or later, our students in the terminal or general education courses do not have this opportunity.

---

[13]In the paper "On the Teaching of Physics as One of the Basic Sciences at University Level," UNESCO Seminar on Basic Science Teaching, Rabat, Morocco, December, 1960 (Proceedings in press, with a shorter version to appear in Contemporary Physics). The following paragraphs are based on a preliminary note published in Am. J. Physics, 28, 577-578 (September, 1960).

In order to do justice to our science, to their needs, and to the commitment which I have urged us to make to a coherent conception of culture, we therefore must stress with care these connections in physics courses for these students. I propose that in addition to the sound presentation of physics itself, we inject into such courses results from the study of certain joint areas, as already discussed. I urge that to start with, at least to some extent or at certain points in the physics course, the representation for a specific topic — for example, Newtonian mechanics — be changed from a pyramid to a "constellation" of related topics, roughly as shown for the xy plane in Figure 2. I should like to call it a connective approach to the teaching of physics, for it is a way of showing some of the interrelated complex of connections existing between any one specific field of study in physics (for example, Newtonian mechanics, or thermodynamics, or electromagnetic radiation, or special relativity theory, or nuclear physics) on the one hand, and other fields of study (for example, mathematics, history, philosophy) on the other.

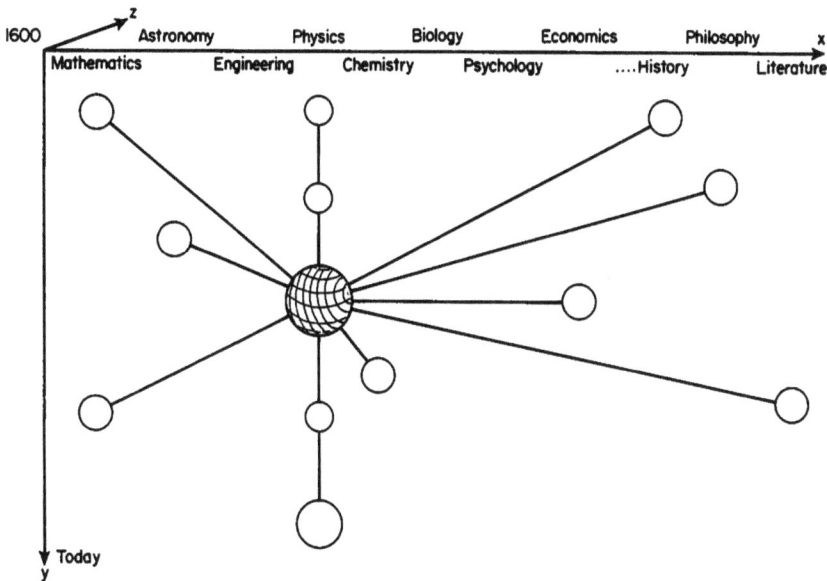

Figure 2. Connective presentation of physics

Therefore, a class in which the special relativity theory has been or is being discussed should also be able to hear, read, or write an essay, on some of the following points (for which we must prepare more teaching aids, monographs, critical bibliographies, and so forth, to help the average, hard-pressed instructor):

1. Galilean relativity (including classical arguments from kinematics and dynamics; meaning of transformation, invariance).
2. Classical concepts of space and time (including Newtonian absolutes, their critique by Ernest Mach).

3. Theories of luminiferous ether (including Descartes, Newton, Young, Fresnel, Fizeau experiment, Michelson experiment).
4. Electromagnetism (including theories of field in the work of Faraday, Maxwell, Lorentz).
5. Einstein's early work (including background, analysis of first relativity papers, critical references to bibliographical material).
6. Other results (including outline of Minkowski representation and summary of, or references to, other sources dealing with cosmology, general relativity — perhaps along the lines of books by Einstein and Infeld or Eddington — and nuclear physics).
7. Epistemology (including Einstein's debt to Hume, Kant, Mach; positivism, operationalism, rationalism; use of the special relativity case to discuss such questions as these: What is the role of a conceptual scheme? Relation of science and common sense. Relation of observation or experiment and theoretical construction. Definition, axiom, and hypothesis in the formulation of laws of science. Criteria of scientific truth. Continuity and discontinuity in the growth of science).
8. Uses and abuses (including spread of "relativistic" views in philosophy and social studies. Effect of "relativistic" ideas on work of art, for example, Durrell. Diffusion and misinterpretation of science in other fields. Ideological reasons for acceptance or rejection of theories, as in the case of the relativity theory in Hitler's Germany).

To repeat, these examples are meant to be not cumulative but suggestive. An instructor would start to transform his course from a traditional physics course to a "connective" physics course by introducing for each of the five or six main topics of the course a few items of this sort, perhaps using no more than 10 per cent of the lecture time plus the assignment of certain reading and essays. While the block on relativity theory would be an excellent opportunity to show connections with philosophy, for other main blocks one could develop the relation of physics to engineering; science and policy making; the role of scientific societies; examples of the use of physics in chemistry, biology, and so forth.

Think now of the opportunities that suggest themselves if in the constellation in Figure 2 the central, dark-shaded area is centered on (instead of relativity), in turn, Newtonian mechanics; thermodynamics; quantum theory; or nuclear physics. Just on the first of these, we immediately can think of the links from Newton's mechanics to the past (for example, the mathematics of the Greeks) or to our present (satellite orbits, particle scattering); to the philosophy of the new-Platonists before Newton and that of Deists after him; to his effect on the models in Dalton's chemistry, and on the sensibilities of the eighteenth century, in which Newton commanded the Muses. Thus, even if only a few references and illustrations are used as we begin to reorientate our general education course, there emerges before the eyes of the student first the intimation and later the proof of physics as a member of a constellation of concerns, so different from the usual, artificial picture of physics as the isolated and stern subject that has nothing to contribute to anything but more physics.

I have, for some years, been working on such a course, indicating or discussing the constellational connections together with the physics content. And to illustrate the variety of possibilities, I think you will be interested in the topics for typical student essays, written after we had studied the block on Newtonian dynamics, in the early part of the year.

For in such a connective general education physics course, one of the several long essays the student is asked to write can be very well coupled with his other main interests (since generally he will not be a science student). In my course at Harvard, the over-all theme assigned to the class for the first major paper has been "On the Impact of the Newtonian Synthesis on X in the Eighteenth Century," where X stands for philosophy or theology or political science or economics or law or psychology or fine arts — whichever is nearest to the interest or the prospective field of concentration of the student. To give better focus to this over-all theme, the student has to define his attention as indicated by a subtitle of his choice, such as "The Role of Newtonian Mechanics in Shaping the Imagination of a Certain Eighteenth Century Philosopher," or whatever. The student is also free to apply for permission to work on another, similar general topic of his choice.

Through his essay the student is again led to engage his own specialized interests, and is asked to couple it to that which represents the major interest of the physical science course. He sees that the course tries to be something specific on its own unyielding terms. And, above all, he is trained to explore the cohesive links that in fact do exist between two fields of scholarship, namely, physics and his own future field of professionalism. I must note here that we find again and again that our students call this essay one of the high points of the course, and that the quality of papers is generally very good.

Here are some of the titles of papers, written by some of the students during the last academic year, on the impact of the achievements of physical science of the seventeenth century on other parts of Western culture:

<u>On Biology:</u>
 "Aspects of Newtonian Science Incorporated by the Leading Eighteenth Century Biologists"; for example, La Mettrie's <u>L'Homme machine</u>, Hoffmann, Boerhaave, Tremblery, Buffon, Lamarck, and some points on which we now see the adherence to a presumed Newtonian model of science misled them.
 "Evolution from Newton to Darwin"; effect of success of mechanist models, and of all-encompassing, nonhierarchical laws.
<u>On Psychology:</u>
 "The English Empirical Psychologists"; the empirical rationalism in Locke's, Berkeley's, and Hume's theories of perception, and their debts to and differences with Newtonian method.
<u>On Economics:</u>
 "Newtonian Influence on Adam Smith"; the search for universal,

natural force and law in economic systems.
"Theories of Causality in Eighteenth-Century Economics"; the
extent to which these were socially and politically determined
despite the attempt to be purely "Scientific."

On Political, Social, and Legal Theory:
"The Political Theory of Thomas Jefferson"; on the acknowledged
debt to seventeenth-century science.
"Henri Saint-Simon and Newtonism"; the development from ardent
Newtonian to violent anti-Newtonian.
"The Effect of Science on Legal Theory", and "Science and the
Commentaries on the Laws of England by Sir William Blackstone";
two fine studies.

On Philosophy and Theology:
A number of interesting essays, including "Bentham's Principles
of Morals and Legislation and its Formal Analogies with Newton's
Principia"; the intentional and unintentional parallels.
"Science and Le Baron d'Holbach". on the most orthodox champion
of unorthodoxy.
"Use of Mechanics in Arguments from Design as a Proof of
God's Existence in Eighteenth-Century Theological Discussions."
"The Opposition to Newtonianism by Berkeley."
"Newton's Epistemology and the Ethics of Kant."
"Seventeenth-Century Mechanics and the Eighteenth-Century
French Materialists."

On Literature and the Arts:
"The Conception of Order in English Poetry before and after
Newton."
"The Effect of Science on the Style and Poetic Themes of
Alexander Pope."
"From Paradise Lost to Essay on Man."
"The Reaction against the Newtonian Synthesis: Blake and
Wordsworth."

I should, finally, anticipate some questions that you will have concerning the proposal of developing Connective Physics Courses.

1. I must re-emphasize that the addition of "external" material need
not and must not detract from the soundness of the scientific content of
the course; the latter is of course the sine qua non of any physics course
given by an instructor whose fundamental loyalty properly belongs to the
science in which he has been trained. In our introductory course for
nonscientists[14] we develop and use the elements of the calculus for all

---

14It has in the Harvard College Catalogue the designation "Natural Sciences 2: Foundation of Modern Physical Science", and is mostly elected
by non-science concentrators in the first or second year, and by pre-
medical students for whom it counts in lieu of a traditional one-year
physics course. The main topics covered are: Galilean and Newtonian
Mechanics; Dynamics of the Solar System; Conservation of Mass and
Energy; Origins of the Atomic Theory in Physics and Chemistry; Waves
and Fields; Quantum Theory of Light and Matter; the Nature of Elementary Particles; Relativity Theory.

students; those who already have some knowledge of the calculus can enter sections in which the calculus is used extensively from the beginning. This is decidedly not a course "about" physics rather than in physics. And I am convinced that this is possible regardless of where such a course is given, because it depends primarily on the attitude of the instructor and only secondarily on the quality of the student. Moreover, one must not neglect the effect of upgrading student preparation for a serious physical science course; what we know to be possible with relatively unprepared students will be that much more possible with well-prepared ones.

2. I have no illusion that such a course will ever be easy to teach, or necessarily will delight every student. The work I propose is not easy. Of course not! Most easy things seem to have been tried already; and in physics instruction they clearly have not worked well enough. The instructor of this new course must build up his material patiently over a long period as he fashions his Connective Physics Course while teaching it. He must read widely, collect respected bibliographies — and learn new things always. His teaching assistants have to be especially trained — though our experience has been that this course attracts the most knowledgeable and loyal assistants, to whom this experience is as fresh and important as it is to the students themselves. Above all, the instructor must really be committed to his approach — otherwise he should not try it.

For the preparation of illustrative material or for some essay bibliographies and advice on correction of some essays, the instructor at the beginning should be prepared to seek advice and assistance from faculty members who are not members of his own department, and who may be scholars in economics or literature, and the like. But in what sense are we members of a "university," of a cultural "community," if this is not done simply? More positively: I should hope that the teaching of a Connective Physics Course helps to cement the bonds of colleagueship across departmental frontiers, so that, for example, there would arise at the college a "shop club" composed of scholars from different areas who share this interest.

3. Even so, local talents should be supplemented by more generally available teaching aids — as I suggested, critical bibliographies, new brief monographs, and so on. I was glad to find that the approach and needs that I have outlined here were found congenial also by the members of the National Commission on College Physics in the U.S.A., and that as a result I have hopes that there will be organized a more or less substantial effort to write, collect, and supply such teaching aids on a larger scale than a single college can hope to provide.

But this approach offers problems and challenges of a magnitude that makes it most appropriate to get international cooperation. It is for this reason that I am particularly glad to be addressing this meeting of the Commission on Physics Education of the IUPAP. If there is sufficient interest in the plan, I propose the formation of working groups that will undertake, by correspondence and meetings, to prepare the teaching aids that are needed for several of the possible constellation topics. A number of groups in different countries can be formed; a regional distribution of tasks can well be expected to produce materials between this congress and the next one. Funds should not be difficult to find, nor interest among the best physicists and other scholars. Thus, for example, I. I. Rabi in his Compton Lecture of 1962 at M.I.T., on "The Education of a Western Man," spoke also of the collaboration of a range of scientists and scholars to write books for science instruction, and he added: "It seems to me, after considering this subject for a long time, that what we really need is interpenetration (of subject matter), and there are signs of a start." He calls for a science text that "brings in other more general ideas and tries to make out of this a humanity rather than something one learns by (rote) or just a system of ideas which would be useful for working problems, but to bring it into a stronger, a better, a human connection without losing in any way the real scientific values and the real scientific rigor which one needs (for study of) that sort. These ideas have been taken up in various places, so I think we begin to see and I am sure that there will be more of it."[15]

If this view of the function of physics instruction in general education can be made generally workable, in many colleges and in many lands, we shall have done justice both to the warning of Miss Mead to keep advanced vocabulary and beginning students in mutual contact, and also to the disguised plea of Koestler not to condemn bright men and women to permanent ignorance of the multifaceted reality seen by modern science. And if we are successful, we shall certainly have produced more than a mere educational improvement. For a curriculum that stresses the connective elements will quite probably be copied from physics (as the successes of physics only often are). This would permit a truly exciting new view of education, in which each student, in addition to penetration in depth of his own chosen field of specialization, will see the rest of the field of knowledge crisscrossed by connective links, with the same major elements appearing as members of different constellations. Thus, even as the Connective Physics Course, when centering on Newton's work, also explores the link to Voltaire and Alexander Pope, so should the Philosophy or Literature course, when centering on Voltaire or Pope, be exploring the link to Newton from its own vantage point.

Some of this is of course already done, but by and large each department of learning presents its course now in an isolated, striplike fashion analogous to Figure 1 — the unfortunate result of too early and too single-minded specialization. It will be when many fields follow our lead and adopt a method sketched in the discussion of Figure 2 that our culture will be seen, by teachers as well as students, to have the coherence which indeed already exists, but which so far has not been nurtured, conveyed, and championed enough in our time. Let us begin here.

---

[15]See also the similar recommendations in Robert Hoopes (Ed.), Education in Science for the Undergraduate Non-Science Concentrator, Report of a Conference sponsored by N.S.F. and Oakland University, Michigan, May 1962 (in press), particularly pp. 175-183 of mimeographed edition.

4

The Design of
Physics Courses

A paper by Dr. C. A. Michels-Veraart analzyed the educational and psychological principles that should be the foundations of science courses for children. Professor Eric Rogers of Princeton University, at present directing the physics project of the Nuffield Foundation in the United Kingdom, gave a paper which elaborated the way in which he approached the construction of an actual course.

SCIENCE IN ELEMENTARY AND SECONDARY EDUCATION*

Dr. C. A. Michels-Veraart (The Netherlands)

My subject is the pedagogy and methodology of the general education of a child, emphasizing the question of the role that physics can, and eventually must, play in this education. The word pedagogy is well coined. It means escorting, guiding the child. It takes account of the fact that the main actor in the process is the child itself. The whole of our task lies in guiding, showing the right way, and in protecting against adverse influences in the development of all the valuable mental and moral abilities the child received at birth. Methodology deals with the methods to be applied for the purpose, and this comes second to pedagogy, notwithstanding the importance of appropriate choice of methods.

A young child can be compared with a tourist who is going to travel in a completely unknown country with the intention, in the long run, of settling there for life. We, as guides, must show him the road, the attractions of the country, the life of the people, their buildings, and their works of art. But much more than that, we have to acquaint him with how the inhabitants think; we must gradually teach him to understand

---

*Material presented by the author has been subject to extensive editorial revision.

their language, their literature, and the expression of their thinking and inner life. At the end of the journey the tourist must be able to understand the new country and its population, to appreciate their way of living, and to join in their conversation. Only if the guide takes all these facets into account does he fulfill his duties. The fact that in our present society this interchange, this mutual appreciation, is so frequently absent, may show that something has been lacking in our guidance.

There is one important point where our analogy breaks down: the guide draws the attention of his tourists to the several aspects of the new country, but the initiative comes more from the guide than from the tourist. In education it must be the other way around. Primarily, observation by the child and his impressions already acquired in an earlier period should form the basis of the discussion between the child and the escort. Certainly in the beginning, education is a play of question and answer. The child has the initiative. What puzzles him, strikes and excites him, starts the argument rolling. The older the child grows, the more our analogy carries weight.

The discovery of causality and the formation of a general concept by generalization are among the first acts of abstract thinking. The child observes that when a door slams it makes a noise. This phenomenon repeats itself time and time again. Many a child is by this observation induced to carry out one of his first experiments: to the annoyance of his parents, he likes to slam a door. From the repetition of the two observations, the child concludes there is a relation. The more frequently he observes that a special event is always followed by the same second event, the more the idea of causality takes shape in his imagination. After all, he follows a similar line to that of an experimental physicist. In the long run he tries to find a reason for everything he observes. The number of questions he puts to everybody he trusts is increasing daily, the "why's" and the "what's" form a big section of his daily vocabulary. Where does the rain come from? How is rain made in the cloud? Why does snow disappear? How does the lamp make light?

In the beginning the understanding of causality is not clear. The logical reason, a result of repeated observation, is mixed with the pragmatic reason. Rain falls so that the flowers can grow; the clouds come so that it can rain; the sun shines so that we can talk a walk. The examples given bear a slight touch of anthropomorphism.

Gradually the period in which the child accepts the idea of animation of objects and is satisfied with an explanation that an event happens because of its results fades away at the age of seven. Gradually the analysis of details takes the upper hand. The observation of detail frequently puts the grown-up to shame, but it is only at about the age of ten that critical realism becomes pronounced, the analysis becomes more critical, and the law of causality takes more and more possession of the child's thinking. This is the period in which he develops logical thinking, in which he begins to arrange his observations in a chronological and logical order. The time sequence is the first to be observed. "When I turn the switch, the light goes on," is an example. Causality follows soon and gradually achieves more importance. At the age of ten or eleven the child becomes interested not only in facts but in relations between facts; he starts to investiage what happens around him. He can distinguish between the main events and the subsidiary ones. Power of observation sharpens, the mind becomes actively inclined to look for and draw conclusions. If physics is to be taught in elementary school, this is the time to make a beginning.

In this development, parents and teachers have to watch how the mental and intellectual powers grow. In the beginning, certainly, the initiative lies with the child. However, the older the child grows, the more the teacher plays an active role. He directs the attention of his pupil to new observations, interests him in events not seen or realized before. In short, carefully preserving curiosity, he makes him see interrelations of effects, the discovery of new laws, and a logical description of the observed facts.

Our task is to discuss what contribution to general education physics can give. Modern statistics can suggest an answer. Allow me to quote a few. Questions put by children at the age of ten are 65 per cent connected with physics. Children between eight and eleven ask twice as many things from the domain of physics as from zoology and six times as many as from botany. Spontaneous questions from pupils in the age limits between ten and fourteen refer to physics 26 per cent, biology 15 per cent, geography 10 per cent, and history 10 per cent of the time. From these data it can be concluded that interest in physics problems starts first, and that later, although physics still predominates, there is a tendency to a more even spread.

Modern education rejects, certainly in the first years, a division of the child's interest into different subjects. Among the different methods which have been suggested, the one that appeals to us most and of which we have personal experience is known as "totality education."

It starts from the environment familiar to the child, for example, the kitchen, our family; for children who live in the country, a farm, a village fair. In the first year one subject can, on the average, occupy the attention for one week, while this period may be extended to three weeks in the fourth year. The intention of this approach is to bring the children to observe and realize everything present in the environment under discussion, to absorb it mentally and emotionally. It starts with a discussion between child and teacher. The child is invited to tell what objects there are in the environment, their number (introduction to counting and arithmetic), their names (written on the blackboard, introduction to writing and reading), what they are used for, what happens in the environment, what is beautiful, what is unpleasant, and so on.

In the discussion with the teacher the impressions are arranged in logical order, so that the whole situation is experienced mentally and emotionally. The pupils are afterwards invited to describe this unified picture, first in telling about it in class, later on in script, drawing or painting, modeling in sand or clay, even in musical form or ballet. This work may take about one third of the available weekly time; the rest of the time is devoted mainly to the techniques (reading, arithmetic, writing) where the basic examples are directly taken from the subjects of the general discussion.

Everywhere a child comes across the results of physics. What does mother use electrical appliances for? How do they work? What must you do to make light when it gets dark in the evening? (Maybe with a slight deviation of thought: why does it get dark?) So the foundation is already laid, waiting for the questions to come later on ... how does it work and why?

Gradually more and more time and attention are given to the different subjects, which begin to take a wider, more general character, a character not so narrowly connected with the immediate neighbourhood in which the child lives.

Contrary to earlier practice, where a subject was chosen by the teachers and treated as a combined effort by the whole class, the individuality of the child is now given more freedom. The children can work alone or in groups; even the initiative of what to take up can come from the pupil. In any case the planning is a combined effort of pupil and teacher. Account can be taken of the level of development of the individual child and his interest. This approach is called "project" work. Physics fits nicely in the scheme followed by modern education and can occupy an active place. Observation from daily life, modern equipment, tele- phone radio, television, household appliances, even impressions received from reading the daily paper, can offer a stimulus. Moreover, besides its direct stimulus to observation and to the search for logical relations, it provides material for an exact description, the choice of the right word or name, the discovery of new words and their exact meaning, new concepts like velocity, attraction, and so on.

An incidental event can suddenly take possession of the child's interest. His excitement forces him to talk about it; the teacher can take advantage of this. The rocket that hit the moon or the other that provided information about Venus can lead to a short discussion on the satellites of the sun. A very elementary account of the cosmos can be given. For the bright children an introduction to centrifugal force may be possible.

The impression may have been created that the subjects for physics are to be taken at random, on the spur of the moment. This is certainly not our suggestion. Different systems are available with a logical sequence, but based on the interest and the level of the pupils. However, they must leave complete freedom to the teacher to deviate from the curriculum if circumstances require it, or if a surprising event happens that draws special attention and can be used to keep the pupil alive to what happens around him in the world.

At the age of twelve the child enters a period in which a firmer foundation must be laid for the intellectual, mental, and aesthetic activities of his adult life. From now on his teacher can play a more active role, gradually taking the initiative and exposing the children to the problems around them. He has to cultivate and bring to bloom the seeds that sprouted to active life in the previous years.

In many countries the child goes for this second period into another atmosphere, and enters a secondary school. Frequently the child no longer stays under the guidance of one teacher but suddenly comes into the hands of a group of specialists. We ourselves deplore this. The unity of education is disrupted; many specialists do not see the child they must educate, but their main intention, their main interest, is directed toward their own specialty. This is a great threat to the real target of general education.

At the end of the four years allotted, nature must stand out for the student as a unity. Nature is not divided into mechanics, heat, optics, sound, electricity, and so on. Any curriculum must start with a realistic picture; transition to the abstract ideas must be gradual. None of these conditions are contradictory. They support each other.

As it is not our intention to devise a specific curriculum, we must demonstrate the idea with an example. We reject the idea of starting in the classical way with mechanics. The concepts of force, energy, and momentum are too abstract for a child of twelve and lead easily to memorization and, for this period, to silly numerical calculations on pulleys, levers, cannon-ball trajectories and objects sliding along a slope with or without friction. Besides these objections, it is for the average child at this age a dull and uninteresting subject. Once the child has received the conviction "physics is dull," his interest is lost for the rest of his life.

We suggest a start with the concept of molecules and their ever-present movement; a concept of velocity can be given in this part of the course. The introduction of collisions — if one particle loses speed in the collision the other gains — is not difficult to convey. The idea of chance can be introduced, and the origin of pressure as the effect of the bombardment of the molecules will easily be accepted. Twice as many molecules in the same volume giving twice the number of collisions on the wall leads to the formulation of Boyle's law, while the game of chance leads to the conclusion that on the average the same number of molecules hits each of the walls per second, thus to Pascal's law. That during the whole process something is conserved will be obvious to the child, but the definition of momentum and energy must be reserved for later.

If I had to give the course, I should switch over now to the movement of electrons in a piece of metal. Here the analogy can be given between pressure and voltage. Meanwhile, vibration would receive attention, vibration of a rod and the first notice of sound as a phenomenon connected with vibration.

I should then return to the molecules, repeat the experiments demonstrating Boyle's law with a gas like butane to find that Boyle's law does not hold. This is the moment to search for a solution and introduce the idea of sufficient accuracy. Let the demonstration with a practical ideal gas be repeated several times with different pupils as demonstrators and so develop the idea of random error and personal error. Again back to butane, show the deviation to be outside the random error. This is an experiment with a surprise effect. There must be a reason. The teacher and the children have to revise their first statement; this is the first introduction of the necessity for scientific doubt and honesty. Attraction between the molecules can be given on trust as the explanation, leading to a lower density near to the wall, resulting in less collisions on the wall, hence lower pressure. It may be now the time to introduce Gay-Lussac's law. The experiment shows an increase of pressure with increase of temperature. More, or heavier collisions, or both are suggested on trust by the teacher. The molecules run faster. The first connection between temperature and molecular velocity is laid.

A trial is now made to see whether a similar law holds for the electron cloud in a metal; the negative answer shows that the analogy breaks down. The reason cannot yet be given but the lesson is learned to be careful with the extrapolation of an analogy, and curiosity is raised.

Returning to vibration, the time is ripe for demonstration with the ripple tank, leading to the concept of phase and intensity and the phenomenon of interference — a preliminary to light as vibration. In this way, always keeping an eye on what the children can grasp, both the corpuscular and the wave concepts are developed in parallel until at the end of the four years' period the equivalency of the corpuscular and wave description demonstrates the unity of nature.

So much has been written about the best curriculum in physics for general education that we have no desire to present another one of our own invention. Rather, we endorse the one suggested at the first International Conference on Physics Education, UNESCO HOUSE, Paris, in 1960, as published in paper 40.[1]

We do not feel we can finish without adding a few words on what must happen with those who, after the four years of general education continue school after the age of sixteen, either with the intention of enrolling two years later at a university or of going out into active life. Must general education stop?

We are convinced that it must not. In these two years we have to deal with two different groups. The older they grow, the greater the danger of a split in mutual appreciation, the danger that the seeds sown in the first years will be strangled by the present attitude of society. Therefore, the more necessary it is to provide means that mutual understanding is preserved. For the so-called humanistic line we should advocate a reduced course in physics, specially directed to the cultural side of physics. We should like to recommend courses in the philosophy of the sciences, the historical development with the impact of the sciences on the structural, economic, and social side of the latest discoveries.

For the science specialist we should insist on a persistent emphasis on the cultural aspect of science, with technology introduced only when examples of application are needed. Furthermore, extra emphasis should be placed on logical, clear thinking and expression and, outside their own field, in reading classics of literature and studying important new works of art.

Physics must be included in general education, not primarily for its own merit and certainly not for its pragmatic value, but primarily because it contributes so largely in educating children to be balanced individuals with an interest in the whole world around them, children who appreciate all the aspects of society, whether these aspects are part of their specialized interest or not, and with respect for the achievements of those of their fellows whose daily tasks lie in other fields.

---

[1]The Teaching of Physics in Schools, Norman Clarke (Paris, OECD). See also International Education in Physics, Sanborn C. Brown and Norman Clarke, (Eds.), Technology Press and John Wiley & Sons, Inc., New York, 1960, pp. 15-21.

TEACHING PHYSICS FOR UNDERSTANDING

IN GENERAL EDUCATION*

Professor Eric M. Rogers (U.K.)

What I have to say concerns the student who finishes his physics at
age sixteen in high school, or at age eighteen or even twenty in a
college — depending on his country or the educational scheme he is
in — but in any case the student who is not going to be a professional
physicist. I am thinking about him ten years later when he is a banker,
or a lawyer, or perhaps a teacher.

I want to ask some questions about those general students.

"What is our purpose in physics teaching?"
"Can we make people more accurate?"
"Students ask me for the answers to the problems I set them.
Should I give the answers? Will that encourage them to think and
remember? Why don't they also ask for the answers when I give
them a crossword puzzle?"
"Are the laws of physics true?"

More important still,

"What is 'being scientific'? What do we mean by 'scientific
method'?"

and

"What are our best aims, when we think of our students ten or
twenty years later, as educated nonscientists doing important
jobs?"

I hope you will think about these questions. They are the questions
with which I begin to teach young teachers by encouraging them to
think out their aims for themselves.

Aims control methods, and methods control syllabi. Syllabi can be
chosen to promote methods, and methods can be chosen to promote
aims. We might aim at giving the general student a well-disciplined
training in scientific techniques with comprehensive knowledge of
principles and definitions. That has been tried, and seems to produce
educated nonscientists who say openly they did not like physics and even
boast that they do not understand it. Nor does that training in scientific
method seem to make such people themselves scientific in behavior. In
fact, psychologists warn us that careful training in some piece of
knowledge (for example, in accurate weighing, or in scientific method)
does not "transfer" to other fields of knowledge or to life in general.

---

*Material presented by the author has been subject to extensive editorial
revision.

Or rather, the student does not <u>often</u> transfer the training. He does not often profit from it in <u>general education</u> — he gains only some specific training. Fortunately there is some transfer — otherwise general education in liberal studies would be worthless. Training does transfer when the student forms a strong wish to <u>generalize</u> and use the training in other ways.

With that warning about transfer in mind, we must choose modest aims for science teaching, if we wish to be realistic and hope for results that will be visible and lasting. We must not hope to train our nonscientists to be scientific, with a full knowledge and practice of some mysteriously ideal "scientific method" such as that artificial scheme set forth by Sir Francis Bacon, and still preached by philosophers but not practiced by real physicists! We shall have little hope of finding our nonscientists leaving school with a good understanding of science <u>if</u> we fill them up with information and tell them we must get it back in examinations in identical wording; or <u>if</u> we drag them through artificial calculations based on memorized formulae; or <u>if</u> we train them logically. A physics course that forms a rigorous foundation for subsequent training of professional physicists will not send our nonscientists out into life with good understanding. For generations we have tried that; and our bankers, senators, admirals, and businessmen suggest, by their attitude, that it is not successful. Instead, I suggest we should teach for understanding in limited areas of physics, so that our nonscientist students develop a cogent feeling of clear understanding, of expert knowledge, of sharing real wisdom with scientists. I am thinking about our young people at a later age, not when they are learning physics, but ten or twenty years later when they are out in the world doing other work than science. They will have to work with scientists, employ scientists, make decisions about scientists, talk to their children about science, and they will live in an intellectual environment where science plays a very important philosophical part. Ten years after school or university, nonscientists will not remember the facts of physics clearly; but if they understood science they will retain some sympathetic understanding. And they will be able to read more science on their own. Even future medical men need to understand the physics that they learn more than they need to accumulate a large vocabulary of definitions, principles, and factual knowledge. Above all, future teachers need a strong feeling of understanding.

Much of what I am saying here should also apply to the future engineer and to the future technologist. A technologist is a scientist of the highest quality who turns his knowledge and skill toward new invention and industrial development rather than to continuing scientific research. A technologist, however, is sterile: one generation of technologists, however good, does not breed a next generation — they must be trained and inspired by pure scientists, so that they have not only a fund of knowledge but a genuine understanding. In our physics we must choose fields that link together, so that things which the student knows and understands in one field are used in another field of physics that he studies; and some things he meets in later fields should illuminate some things he met in earlier fields. Essentially, there must be time for students to use these chosen pieces of physics. In that sense, the choice of topics for a syllabus is important — the topics should link together to show a growing fabric of knowledge. On the other hand, I do not think there is any single scheme of topics that is better than all others. Any set of topics, provided it is not overfull but gives room for discussion and teaching to give understanding, is a good one if the teacher can teach it with goodwill with his aim set on giving understanding.

How can we do this? I can make only suggestions; but there are things that we think we have found successful over the past fifteen years. First, teach less material. Omit some pieces of physics, so that the syllabus is not too crowded, so that there is time to teach for understanding. Then students will emerge ready and able to read any of the missing material that they wish. For the things we do teach, we should choose pieces of physics that have many uses. I do not mean just practical applications but rather linkages with other pieces. Physics should appear to students as a growing fabric of knowledge in which one piece that they learn reacts with other pieces to build fuller knowledge. We must be very careful to introduce any piece of physics that we teach with clear indication of our purpose, saying clearly how we are trying to build more physics. And after we have taught a piece of physics, we must look back on it and talk with our students about the way in which that piece of physics fits in with the rest of physics and builds more.

We must not just give formal definitions to be memorized or statements of principles or laws to be used mechanically; that would be asking our students to behave as a rubber stamp, to reprint on every examination paper the standard things that we have taught them. Medical students who seem to need the formal facts of physics do not actually remember them clearly ten years later. What they need instead is an understanding of the way physics is done, illuminated by careful study of some samples of physics. Even professional physicists will go faster and farther if they start in school with the feeling that they understand physics.

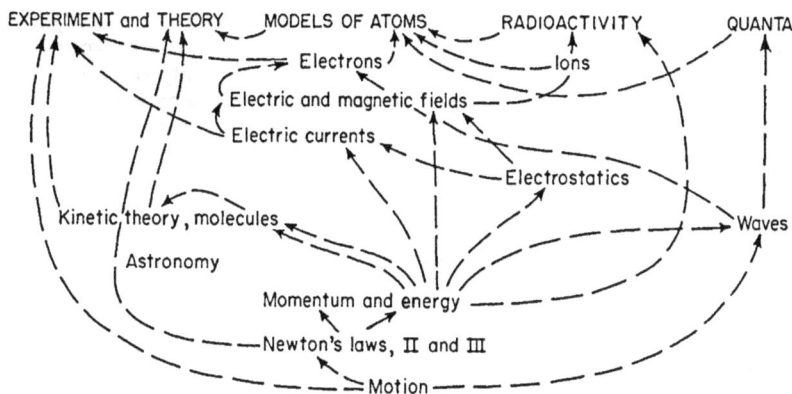

Figure 1. Example of syllabus

As an example, I will show how one can construct one suitable syllabus, such as shown in Figure 1. Note that I construct it backwards, beginning with the end points, my ultimate aims for my students; and then finding out what earlier topics seem to be needed to support those aims and to provide the groundwork for understand the later teaching.

This is just a syllabus that I myself happen to like and use. Another teacher should make his version, of course. And the most important thing is to gather would-be teachers together and have them construct a syllabus like this. Of course there will be outcries over the things left out. In the notes in Figure 1 I have omitted hydrostatics, statics, geometrical optics, and Wheatstone's bridge, and so forth. These may be important, but it is essential not to overcrowd the syllabus or we defeat our own object.

Note that I have chosen to end with some modern physics, and I hope students will emerge with some understanding of "models" and theory and its relationship with experiment. I have many other minor objectives such as acquaintance with interpolation, extrapolation as a road to great discoveries, the use of rough approximations, and so forth — those are dealt with by the wise teacher whenever he sees a good opportunity.

When we have made such a syllabus, the battle is not won: an efficient teacher can still ruin our hopes by dictating formal information and giving detailed instructions in the laboratory so that he does the experiment for the children. We need to guide the teacher toward a different frame of mind, asking himself what each demonstration shows, how it fits into our knowledge of science, whether it provides absolute truth or only some illumination, whether the result that emerges is a statistical average or a unique quantity.

All these are matters that we must have would-be teachers go through by constructive, critical argument themselves — we must teach them in the manner that we want them to use for teaching students, and with the same aims.

We need to offer our students experiments to do on their own in the laboratory. That need not cost much money. They need simple apparatus, they need a silent teacher, they need some encouragement from the teacher, they need some questions about their experiment from the teacher. And, above all, students need plenty of TIME to do their own experiment and make mistakes and enjoy successes, and then to think and argue about their own observations in the experiment.

For such experiments we should not always choose especially complicated or even especially modern apparatus. We may give each student a simple spiral spring (wound from piano wire, costing a penny or two) and ask them what they can find out. Left alone for one hour they will "discover" Hooke's law. Given another hour, nothing much more will happen. Given several more hours (in the course of a week or two), much ingenious experimenting will emerge. Another example: we can teach a lot about electric currents by letting students work on their own with simple, robust apparatus. First, a battery and a lamp: light the lamp. Then make the lamp brighter and dimmer; then, insert an ammeter to measure the current. (We give the ammeter as a ready-made instrument, to be accepted as a "black box" just as we all accept a stop watch.) Then, insert a voltmeter. We do not say how to insert the voltmeter — we almost encourage students to put it in series first, by our careless instructions. Within six weeks we have our students using a triode tube as an amplifier: they hear it amplify a musical signal that we provide; and they see the amplification on an oscilloscope — and they understand quite a lot of all this because they have done it themselves.

With good teaching to promote understanding, quiet creative experimental work in laboratory, and critical discussion guided by the teacher, we might hope to promote understanding of physics — and of physicists and the way they work — but we shall fail completely if we ask the wrong kind of questions in homework or in tests or in examinations. It is no good to insist in class on understanding and then ask for formal answers in examinations — a definition of coefficient of cubical expansion to be returned in the exact words of teacher or book; or a mechanics problem to be solved by putting numbers in a memorized formula. Therefore, we must ask teachers to remember their aims at every point in constructing tests; otherwise, students will be guided in the opposite direction when they take the tests, and they will conclude that after all science is a set of formal rigmaroles unconnected with the real world or with clever, sensible thinking. But it is not easy to make sure that all our questions ask for intelligent thinking based on the physics that has been studied. Nor is it easy to remember to insert questions like that in the middle of our teaching so that we make students think out the next stage of their knowledge for themselves.

That is not easy for teachers nor popular with students. Inertia is always with us: memorization is easier for students, perhaps the sole academic activity of a stupid person; and it makes lighter work for the teacher who has to read the things that students write. In Princeton, where I insist on asking questions that require constructive thinking, 10 per cent of my students quit my course at an early stage, saying openly "I had not expected to be asked to think."

We must give teachers encouragement and guidance in making questions that promote thinking, and we must repeatedly insist on the importance of using those in our teaching and in our examinations.

When we have manufactured questions that point toward our aims — questions that encourage students to think constructively and put together several pieces of physics they have learned — we still have to read and mark, sympathetically, the answers that students give in homework, or tests, or examinations. For that we must adopt the attitude of one scientist talking to another, albeit in simple language. I have in mind the kind of talk that one hears in a research room when neighboring scientists come in and stimulate the research man with critical comments, irritate him with bright, helpful suggestions, or even waste his time by exchanging ingenious questions about physics with him. We must not insist on the student's answer taking a particular form or even being the particular physical reason that we expect. We must reward every piece of intelligent thinking, though we should punish stupid answers or lazy, or antiscientific ones.

I find we can redirect intelligent, willing teachers so that they teach for understanding; and, when they do that, they will find that their pupils say they do remember and use some of the physics which they, in turn, learn and understand. We must persuade those teachers to be content with teaching less physics; but we can assure them that in teaching for understanding they will need to use many well-linked pieces of physics and teach those pieces so thoroughly that they will still be teaching a large quantity of facts and principles of physics.

Even with new teaching aims and methods, I do not hope for great results to spread quickly across the world, or even through a group of students. In such an important change we must expect only a slow growth: man takes long to change his mind in the higher values, as every great religious teacher knows. A major influence that sets the stage for science teaching, before school starts it, is the attitude of parents toward science in the home. So, I hope that all teachers will remember that in teaching science to children they are laying foundations for the attitude toward science that those children will have as parents of the next generation.

APPENDIX 1

SOME EXAMINATION QUESTIONS

First, I ask you to note the following:

(1) I do not use "objective tests" with 5 suggested answers to choose among, because I think these turn the examination into an intelligence test; and they are apt to stultify the physics. Instead, I offer space — just a few lines — on the examination paper for the student to write his own answer. Then a physicist, but not a machine, can judge him. Where that is intended, I have indicated the length thus "(4 lines)", meaning that on the examination paper we provide space for 4 lines of writing.

(2) Remember that questions which require several pieces of physics to be put together constructively are dependent on our choice of syllabus. Therefore, a question that is a very good one in one course may be a bad one for another course, which does not provide the student with the necessary factual background.

(3) These are only a few scattered examples. To gain a clear picture of our policy, one needs to look at the syllabus and its treatment and then look at a large set of test questions. More sample questions are given in my <u>thin</u> book "<u>Physics for the Inquiring Mind.</u>"[1]

1. (Near the end of the course when students are familiar with energy and electric fields and have met ions, and so forth)

A "gun" is set up to accelerate positively charged particles from rest to speed v by an accelerating potential difference V volts.

When the gun is used to accelerate protons, they emerge with speed v_p

---

[1]E. M. Rogers, <u>Physics for the Inquiring Mind</u>, Princeton University Press, Princeton, N.J., 1960.

When the gun accelerates "alpha particles" He^{++}, they emerge with speed v_a

(i) The ratio v_a/v_p will be

(ii) Why is our answer to (i) independent of gun voltage V?
(1 line)

(iii) For very large gun voltages the measured ratio would be slightly different from your prediction in (i). Suggest a reason
(2 lines)

(iv) For extremely large voltages the measured ratio is nearly 1.0 Explain:
(2 lines)

2. (For students who have studied simple kinetic theory)

A. A gas in a cylinder with a frictionless piston is suddenly compressed by a man pushing the piston inward. The gas grows hotter.

(i) Describe, in terms of molecular behavior, the mechanism or process by which
the gas grows hotter
(2 lines)

(ii) Where or what is the heat that is gained?
(2 lines)

(iii) Where does the heat that is gained come from? What provides it? (NOTE. The piston grows no cooler.)
(2 lines)

B. (i) A compressed gas in a cylinder with a movable piston is allowed to expand by pushing the piston out.
Explain briefly why the gas cools.
(3 lines)

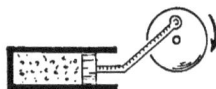

(ii) If the piston is connected to a frictionless
flywheel, what happens to the heat lost by
the gas?
(1 line)

C. A small capsule of compressed gas is placed in a
large bottle from which all air has been pumped out,
so that there is a vacuum. The capsule splits open
and releases the gas. Explain why in this case you
would NOT expect to find the expanded gas any cooler.
(4 lines)

D. Most real gases do show a small cooling when re-
lease as in (C). What does this suggest regarding the
molecules of such real gases?
(HARD. Make an intelligent guess and give a brief reason for it.)
(3 lines)

3. Describe the meaning and use of "laws" in physical science, discussing examples such as Hooke's Law, the Law of Conservation of
Momentum, and so forth.
(Obviously there is no single right anser to the question "What is a
scientific law?" You are invited to give several opinions.)

(Suggested limit: one page).

4. It can be shown by algebra, calculus, or geometry that <u>if</u> acceleration dv/dt has a <u>constant value</u> a, <u>then</u> distance s = $v_0t + at^2/2$. And, for motion starting from rest, this reduces to s = $at^2/2$, or s varies directly as t^2.

An experimenter times the motion of a small truck down a slanting railroad, starting it from rest. He plots the results on a graph of s vs. t^2, and he draws the "best straight line." His points are very close to it. He draws error boxes around them. He then offers each of the following comments:

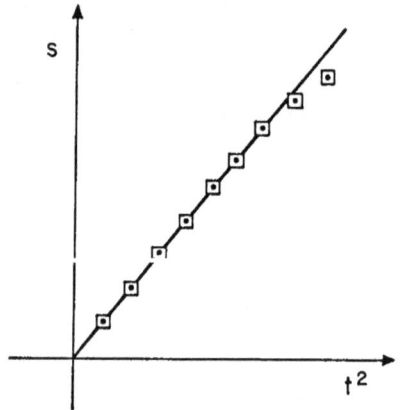

A. <u>In each case explain briefly why the comment is not a wise or sound one.</u>

   (a) "The fact that my points lie close to the line prove that s $\propto t^2$ for motion with uniform accelerations."

                 (2 lines)

   (b) "The fact that most of my points lie so close to the line shows that some of my error boxes are too big. I must reduce them."

                 (1 line)

   (c) "The fact that my points lie close to the line proves that my measurements were very accurate."

                 (2 lines)

   (d) "The fact that the last two points are off the line shows they are bad measurements that should be erased."

                 (3 lines)

B. Suggest a reason (other than chance errors or crooked rails) for the fact that the last two points fall below the line.

                 (3 lines)

5. An experimenter makes a truck accelerate by pulling it with a load hung on a wire over a frictionless pulley, as in the sketch below.

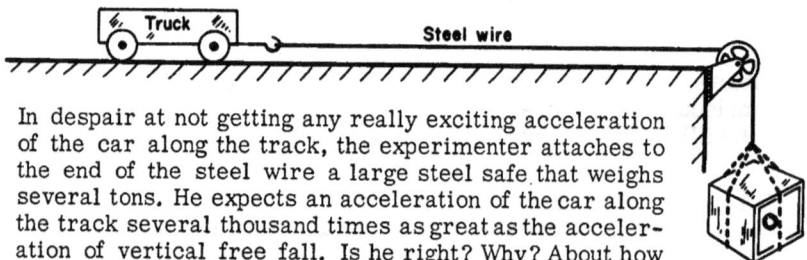

In despair at not getting any really exciting acceleration of the car along the track, the experimenter attaches to the end of the steel wire a large steel safe that weighs several tons. He expects an acceleration of the car along the track several thousand times as great as the acceleration of vertical free fall. Is he right? Why? About how much acceleration do <u>you</u> expect? Be careful in wording your reply to <u>be as nearly quantitative</u> as you can be without accurate measurements. For example, you might say (wrongly): "The acceleration will be obtained by dividing the acceleration of gravity by several thousand." Give a clear justification for your answer.

(5 lines)

6. (A long question, deserving 30 minutes, at the end of the course, to be answered on separate sheets of paper.)
What, in your opinion, are the goals and methods of theoretical physics? Illustrate your answer by discussing one or more examples, such as the following (or you may choose other examples, if you like, from physics):

    Rutherford atom picture   Relativity   Quantum theory
    Simple magnetic theory   Kinetic theory of gases

7. (A problem about oil film measurements, at an early stage.)
  (i) Describe an experiment that you have done to estimate the size of an oil molecule.
 (ii) What risks or assumptions did you have to take in arriving at your estimate?
(iii) What does the "size" that you estimated mean? Are you sure it is a true and absolute size?
(iv) From your estimate you can obtain a guess of the size of an atom. What other information do you need for that, and where can you obtain that?

      On discussing Professor Rogers' paper from the point of view of the educational problems of Latin America, Professor F. Cernuschi (Argentina) had this to say:

"I should like to add something to what Professor Rogers said, that one has to teach physics through experiments made by the children themselves and that they must be given adequate time for this. To achieve this we must modify not only the syllabi of the sciences but also the general framework of secondary education. Especially is this so in Latin America, where we have the encyclopaedic approach that drowns the students with data to be memorized. If the syllabi of secondary education are not modified so as to be able to reduce the number of classes, at least by a half, the aims described by Professor Rogers cannot be achieved. The student must be given time to learn how to think and to experiment. This is true not only for physics but for other disciplines as well. That is why I suggest that we must change radically the general objectives of secondary education. We must discard completely this encyclopaedic attitude that does nothing but suffocate the reasoning and observation powers of the students.

"When we consider attempts to unify scientific and humanistic education, we must not forget the lessons of earlier experiments which failed. I have in mind the famous Chicago Plan. The basic idea of this plan was to use as the basis of study the most outstanding masterpieces that had been written in the whole history of mankind. Thus we had Dante, Cervantes, some fundamental works of Galileo, the original works of Newton, Fourier's theory on heat conduction, Maxwell's field theory, Einstein's relativity, and so on. By simply considering the list of books that were selected, one can see that this scheme was not feasible. It seems to me that the authors of the Plan had not even opened these books. One cannot understand the heat conduction theory unless one has had several courses of mathematical analysis. Therefore, the Plan was a utopia and was inevitably doomed to failure. I believe that its failure lay in the fact that it was a one-sided attempt at unification made by philosophers with inadequate scientific knowledge."

5

Experiments in Teaching Physics

THE BRAZILIAN INSTITUTE OF EDUCATION,

SCIENCE AND CULTURE (IBECC)*

Dr. Isaias Raw (Brazil)

Elementary science teaching is facing a dilemma created by the rapid development and widespread importance of science, and its impact on everyday life. The problems of science education are very difficult, especially for underdeveloped countries striving toward development, as Brazil is.

The initial assumption of our project is that it is the function of the scientist to evaluate the impact of science on general education and on daily life. It is the scientist who should review continuously the evolution of science, relate it to the changing conditions of his country, and direct science teaching, from the elementary school on. One cannot rely on officials or general education specialists to decide science curricula in schools, what is no longer important or what is basic to new aspects of science.

Many countries maintain nationally controlled curricula. Brazil has just got rid of this mistake, which for several generations had prevented our educational system even from trying new approaches. One cannot improve science education by merely changing an official curriculum. The problem is to change the minds of teachers and students as well as the opinions of officials of Education Ministeries.

We came to the conclusion that some effort should be made to introduce science teaching closely related to daily life in primary schools. Most of our children do not receive secondary education, and in primary school some mass impact must be provided to prepare the youngster for modern life.

---

*Material presented by the author has been subject to extensive editorial revision. This paper was submitted to the Conference but not formally presented.

Our secondary schools were created, as in many other countries, in order to prepare an elite for the university, to provide the few lawyers, doctors, and engineers necessary at the time. Although secondary education has expanded, it maintains basically the same aims and generates in students and teachers an attitude of contempt toward manual work, including laboratory work.

Our approach was to introduce and promote as much laboratory work as possible; the immediate result was to replace by basic science and a genuine scientific approach the "chalk and talk" methodology that for years had made only slight modifications of the early teaching of science, as imported from Europe through books like the classical Ganot's Physics.

When changing science teaching, the logical way would be to improve the preparation of future teachers. In countries like ours, we cannot afford to wait for such a slow program. An alternative is to begin working on the present teachers, a difficult task, especially in the present conditions in underdeveloped countries, where teachers have not been required to have any initiative, all responsibility being borne by the State through the controlled curriculum.

Our initial effort was concentrated on students by trying to encourage them to do scientific experiments on their own initiative.

Our student program is today made up of the following activities:

1. We supply equipment, at cost price, and instructions for home use by students.
2. We promote interest in science and scientific experiments by television programs, talent search competitions, and science fairs.

We produced initially a chemistry kit. This kit contains equipment and about fifty chemicals, so students could do experiments as directed by a series of booklets. These booklets contained about thirty experiments, beginning with simple ones, cookbook type, and evolving to more complex ones, with less information. At present the Institute produces kits in chemistry, physics, biology, mathematics, and general science.

As these kits cost about twenty-five to thirty dollars, we added a similar but more restricted program: small condensed kits are provided for a number of related experiments. Included is a booklet with more experiments than those possible with the kit.

Science fairs were organized, where students from a certain region present small projects and apparatus they have built. The fairs are open to individual, class, or school presentation, and young people meet and discuss with colleagues their common interest in science. A few prizes are provided and lectures are given by scientists.

Television programs have been tried with good results, presenting experiments with only limited explanation to provoke interest and encourage students to consult books, teachers, or adults. Simple experiments were repeated in a week by thousands of students.

A talent search differs from the science fair by being nationwide, where the best students compete for fellowships, presenting their reports on experimental science or mathematics. The selected students are taken to the annual meeting of Brazilian scientists, where, by interview, the best obtain fellowships. They have the opportunity to observe a real scientific meeting and to discuss their problems with scientists.

This program of providing facilities for student experimentation to raise student interest and to create conditions where they can meet has given good results. Science club type activities were tried with little success.

Our second step was to influence secondary schools. It was done again by a group of activities:

1. Providing simple equipment, at cost price.
2. Providing instructions for using the equipment.
3. Organizing a system of visiting science teachers.
4. Providing Summer Institutes for training teachers.

The secondary school, especially the Colégio (senior high school) has evolved in an attempt to keep pace with science progress. Scientific apparatus has in the past been considered unnecessary; if available, it was shown through windows and never used even by the teacher. Our effort has been to provide the schools, at low price, with simple basic equipment. The students' kits changed the idea of how it is possible with restricted funds to organize a school laboratory in such a way that each student can do the experiments. Today this equipment, at low price, is available to any school.

The equipment is available with detailed instructions for teachers. A monthly news bulletin and other teachers' publications are distributed free. A yearly congress of science education is organized. It is held as part of the annual meeting of the Society for Advancement of Science.

The television program, organized not to replace the teacher but to raise doubts in the students' mind, brought so many questions that it became necessary for teachers to watch it. Another scheme was started by establishing a visiting professor program. The professor visits a school and gives a demonstration lecture with simple, improvised equipment, leaving to the local teachers the task of keeping up to this standard. The results have been good, and local teachers welcomed the visits.

New curricula have now been devised by the cooperation of scientists, University teachers and secondary-school teachers. Teachers are being prepared through Summer Institutes. For teachers who cannot reach the necessary level for the new curricula, more restricted Summer Institutes are held, just to improve their knowledge and teach them simple experiments.

The Universities, especially in the South where the number of candidates is large in relation to available places, have established a very difficult entrance examination. The level of the examination to select the so-called better students is raised from year to year. There are about twenty candidates for each place in schools of medicine and engineering. This entrance examination, prepared by the University, has made the secondary schools cram students to pass this examination, with no regard for what they ought to learn. Today the Institute is providing facilities and equipment to introduce laboratory work and reasonable tests to replace the examinations that ruin the last two years of "colégio."

By involving University personnel in the whole program, scientists and University teachers have become interested in the improvement of science teaching. New curricula are being adapted or created with their help. The changes they are helping to promote in secondary schools, the better preparation of their students, and the improvement of entrance examinations are slowly beginning to produce changes in the University itself.

IBECC, officially the national commission for UNESCO, operates as a foundation. This service is organized by the University of São Paulo. Help has been received from the Rockefeller and Ford Foundations and the Brazilian National Research Council. Further support was provided by the Pan American Union and the National Science Foundation. It operates under contracts with the Ministry of Education and with State and Municipal Education boards.

SOME EUROPEAN DEVELOPMENTS IN PHYSICS TEACHING*

Professor D. Sette (Italy)

I shall report briefly on some efforts at present carried on in Europe under the sponsorship of OECD. This organization has realized that the economic development of its member countries is intimately related to the number and quality of their scientific and technical personnel. And it is therefore stimulating a number of activities designed to improve the situation. Pilot courses in physics have been organized in secondary schools in various countries, among them the Scandinavian countries, Spain, Italy, and Yugoslavia. Almost all of them make use of the PSSC material, at times with some modifications so that the course will correspond better to local requirements. The usefulness of such a project is not limited to the increased use of a modern course. OECD is bringing together the educators in the various countries using the new courses, to discuss the pilot experiment, and it is hoped that new ideas will emerge. OECD is carrying out similar pilot experiments in mathematics, chemistry, and biology, and it is in the program that when the pilot experiments in various sciences have been carried on for some time in a number of member countries, a close collaboration between the various specialists involved would be stimulated in order to coordinate curricula and increase the efficiency and the impact of scientific teaching.

Professor Zacharias has told us that attention is being given in U.S.A. to the teaching of science in elementary schools. OECD is also con-

---

*Material presented by the author has been subject to extensive editorial revision.

sidering how to stimulate and support projects for the teaching of science at this stage. This development is still in the early stages. Nothing at all has yet even been thought of for science teaching in the last years of secondary schools. The problems involved in both the introductory and the conclusive coordinating cycles of science curricula for general education require close collaboration with other disciplines and the development of the methodology of their presentation.

In Europe it is hoped that another international organization could take its responsibility in this direction — the Council of Europe. This organization is sponsoring periodic meetings of the ministers responsible for education in the Western European countries. In the last one held in Rome in the autumn of 1962, the problem of the role of humanities and science in the education of modern man was discussed, and the necessity of balanced education was stressed.

It has frequently been pointed out that it is important to have close contact between schools of different levels and between schools and industry. A system that exists in the U.S.S.R. seems to be quite effective. Every secondary school is intimately associated either with a research and educational institution of higher level or with an industrial establishment. The school is helped in the organization of its science teaching and in the organization of scientific conferences, seminars, and extracurricular activities. The students have easy access to the research laboratories of the particular institutions. The U.S.S.R. authorities seem to be pleased with the way in which the collaboration is working.

A teaching experiment that has shown its value in particular situations of shortages of teachers and equipment is that of mobile units. Experiments have been made in the U.S.A. and recently in Sicily with an OECD project. The latter project will be extended to Greece and Turkey. It consists of a car serving ten small towns. Two well-qualified physics teachers and two technicians form the team. The experimental material for a one-week tour is carried by the car. Classes are given each week in eighteen schools. The system has been devised to improve immediately the courses that can be taught with the limited teaching facilities available in the regions. The results have been very good, not only for the education of the pupils but also for the improvement of the teacher's resources. The system of mobile units can have some application in other undeveloped regions for circulating equipment of high cost and short effective life.

Everybody is familiar with the method of teaching which is called "programmed instruction" and which is based on the findings of behavioral science and experimental psychology. The student's attention is focused on the significant points of the subject study. He is tested on what he has learned, receives confirmation that he has mastered the item, or is asked to go back to the points which explain his error. The advantages claimed are that the instruction proceeds by small steps determined by the student's ability and that it requires active response from the student. Moreover, the process of learning does not require the presence of the teacher but only books and aids of various kinds. The limitations of the method are evident. While it seems immediately useful for teaching factual information and techniques, it does not seem equally fitted for treating disciplines where basic concepts are present and where intuition and fantasy have an important role. We are, however, at the early stage of this technique, and it may be that if talented physicists give attention to the method, something good may develop in the future for the teaching of physics. This seems to me, at present, to be quite remote.

What seems to be more possible is the use of the method where the introduction of basic ideas is not involved. This is the case with the continuing education of professional people. The possibility of applying the method of programmed instruction to the retraining of teachers should be investigated. One must always consider the high cost of programmed instruction in money and time. At present, the cost of producing one student hour is about $1000 and the time required, about 100 manhours. The position is similar to that of printing. The technique can be applied only if large numbers of people are going to use the course.

A final point I should like to make concerns the influence that the teaching of physics may have on the wider use in mathematics of computers. Physicists and engineers do not take full advantage of the possibilities of existing computers, largely because computer science has not been part of their training. There is a movement to introduce computer science into mathematical courses at college level and to let students use computers for the solution of their problems in the advanced courses in universities. This change in mathematical curricula is warmly welcomed by physicists, not only because it will allow a wider use of a tool which is steadily increasing its power, but also because the change will bring into an early stage of the mathematical curriculum subjects of great importance for physics, like the theory of probability. Experiments of this kind must start at college and university level, but very probably they will in the long run influence also the mathematics and physics curricula at a secondary-school level.

CURRICULUM REFORM IN THE U.S.A.*

Professor Jerrold R. Zacharias (U.S.A.)

Three years ago the IUPAP had a meeting in Paris, and at that meeting I reported on the Physical Science Study Committee work on physics for American high schools.[1] There were a number of new

---

*Material presented by the author has been subject to extensive editorial revision.
[1]International Education in Physics, Sanborn C. Brown and Norman Clarke, Eds., The M.I.T. Press, Cambridge, Mass., 1960, Chapter 5.

things in PSSC, and by now they are quite well known. I should like to show what has since happened, not just to PSSC but in the U.S. to the whole problem of curriculum reform, which is to a certain extent orientated according to the PSSC principles, the main one of which is to bring university professors, that is, research workers who understand the subject in detail firsthand, into close contact with teachers in the schools and to generate, from that combination, a new set of working material.

The PSSC is a collection of learning aids, not teaching aids. They are aids to the student, and the teacher is regarded as the most important learning aid. To make such a collection, one must first understand one's subject and decide which parts of it to choose. As Professor Rogers has said, one should always design a curriculum by picking out the end point and working back. We decided that we should like the citizens of the U.S. not to be afraid, for example, of the ideas that underlie the quantum theory. We should like them to understand the whole notion of quantization, the whole notion of particles and waves, and that there is no duality in the particle and wave picture. In order to clarify this, working backwards, we said that it was necessary, clearly, to understand the electrical nature of matter, the atomic nature of matter, and then, of course, working back, Newtonian mechanics. It is also necessary to know why one believes Newtonian mechanics. One believes Newtonian mechanics because of celestial mechanics, not because of blocks of wood on tables. One believes it because of the universality of it and the experimental evidence. So we worked back to form a course appropriate to about a year and a quarter, or a year and a half of an American high school, knowing that it could be broken up in any kind of subdivisions.

The laboratory is integral with the course, and if there were any part, any learning aid, that I should demand having in it is the laboratory, with the students using their hands, their eyes, their ears, their noses, their throats, and their larynxes. But the laboratory must be made with a guide, a book that cuts a fine line between giving directions and being completely free so that the student is completely lost. You must have a partial map in the laboratory guide. I think that this has been done.

Now, films.[2] There are any number of reasons for using them, but one most important thing. No film, no television that I know will substitute for being able to get your hands directly on the thing. Although we had to work very hard to learn how to make films, films are not enough, but neither is anything else enough. Neither is the teacher!

For the teacher, of course, we had to make a guide that puts all this together, and the guide is 10 centimeters thick; the textbook is only 3-1/2 centimeters thick.

We also have teacher training. Fortunately, the National Science Foundation was able to support our teacher training institutes, and so far almost 4,000 teachers have gone through these summer training institutes in the U.S.A.

---

[2]The PSSC films shown during the conference included "Time Dilation," "Matter Waves," and "Vorticity."

What has happened since that Paris Conference? Three years ago, I said we started with 300 students, just with mimeographed material. We then made a jump to 300 teachers and 12,000 students, the next year to 24,000. Since that Conference the student numbers have gone 43,000, 80,000, and this year 130,000. That is about 30 per cent of all students taking physics in the U.S. high schools. It includes every type of school. The PSSC course is very heavily adopted by the Catholic parochial schools in the U.S.; it is poorly adopted in New York. In the state of Florida, there is now 90 to 100 per cent PSSC. In Florida the course went not quite as well in the all-Negro schools as in the all-white schools, because neither the teachers nor the students were as well trained to begin with. I do not know the numbers of PSSC schools or students throughout the world, but there have been many adoptions. The big countries of Germany, England, France, and Russia are paying no attention to it. There is no German translation; there is a French translation, but it is Canadian-French. There is a heavy adoption in Canada. The Scandinavian countries are helping us to modify it to fit their needs, so there will be translations in Swedish, Danish, and Norwegian. I have seen some Italian, Spanish, and Portuguese translations. There are translations into Hebrew; there is a very good Japanese edition but no Chinese; there is a Turkish and a Thai (Siamese) edition in preparation. All these are translations mostly of texts, laboratory guide, and, in some cases, teacher's guide. We have not yet translated but we should be translating films.

We decided at the very beginning to write little books of 150 to 200 pages to show how physics enters into all sorts of subjects. Thirty have been published so far and translated into 15 to 20 languages.

We have now decided, using the philosophy of PSSC, to make the course appropriate to the American college, what the British would call the sixth form. The advanced topics already include good sections, with all learning aids, on Special Relativity and on Angular Momentum. Work is now being done on a detailed energetics of atoms and molecules, so that one can have a course that, for the citizen of the U.S., will make a solid course in science. It must be made so that it does not frighten him, or so that it does not frighten her, which is even more important. (We inverted the order of the PSSC and started essentially with waves and optics instead of with Newtonian mechanics — cold-bloodedly — in order not to lose the girls! And that has worked).

The PSSC was made so as to fit into the middle of an educational system. We had to make the beginning of the PSSC in such a way that the first quarter of it could move down into earlier years. In the U.S. we call this the ninth grade. We have a group working on teaching elementary-school science, again with a complete collection of learning aids, using a little more chemistry and getting in some important scientific notions. From a project concerning African education we learned that we did not know anything about what it means to teach elementary science. Francis Freedman pointed out that "Every professional science book rests on the philosophy of starting with the known and going to the unknown. And this is the surest way to bore the children to death. Start with the unknown and work to the known." Let me give you some examples of how we are starting from the unknown and going to the known. One nice experiment is about a simple pendulum. A simple pendulum is simple. If you start a student, a child of seven, eight, or nine, with a pendulum swinging back and forth, he says "Yes, I am accustomed to that. That is not very interesting." However, and we have tried this with students, start with the double pendulum, two pendulums coupled. As every physicist knows, with two coupled pendulums you have lovely phenomena. You get the students so involved in this intriguing thing of two pendulums that soon they want to know about one pendulum. And this happens every time.

Another example is this. Children do not care whether something is physics, or biology, or chemistry, or mathematics. It is nature. It is only those of us who work in the distilled atmosphere of the university who insist that there be physics, or mathematics, or biology. Children do not care. So we have two months of science effort on biological cells. They start by using a microscope. (The Japanese came to the rescue on this point. One can buy an adequate microscope in the U.S.A. for $5, imported from Japan.) The first thing the student must understand before he looks at, say, the tip of an onion root, to see a cell, is the meaning of scale. How big is it? You can get the children into the kind of scientific trouble that a man should be in, and in addition you teach him something of the chemistry of the cell, the biology, and so on.

I have said nothing so far about what happens beyond PSSC, the college problem. A number of universities, some separate and some together, are trying to follow the PSSC and to make a course that does follow it. We have a lot of detailed work going on at M. I. T.; we have a science teaching center there, separate from the physics department but associated with it, for making the learning aids that will go with the course for scientists and engineers to follow PSSC. Other colleges, other universities, are working up a two-year sequence.

In the U.S. we have about 180 million people, as many people as you have in Latin America. We have universal education. Without education the curse of superstition and prejudice cannot be changed. We have in the U.S. one million and a half teachers. We have a rate of dropping out of our teaching profession of about 20 per cent per year, mostly women who go off to be married young, and mostly from the early grades. Therefore we have a teacher retraining problem and a problem of training new teachers, which is just as serious as the problem anywhere. We must learn how to teach teachers to handle materials that depend on knowing why you believe what you believe. This means that in the U.S. we should be training and retraining about 300,000 teachers per year. In about ten years we would just about catch up. This implies that we must learn how to make the kind of educational technology that we use in PSSC, work for the training of teachers. The only way, I am convinced, is to have the materials in a form which is refractory, which cannot be changed easily. That means there have to be very careful choices of subjects, of texts, of words, pictures, laboratory, films, and readings.

I have made some cost estimates, and I think that with a few million dollars per year we can really make headway in this problem. We, at this time, are spending 30 billion dollars per year on education. Some of the rich cities are spending $600 per student, some of the poor states, like Kentucky or Mississippi, are spending $250 per pupil. $30 billion is a big industry. The automobile industry and the oil industry of the U.S. add up to about the education industry of the U.S. The people in our Congress and the people in the executive office, in fact our President Kennedy, understand this perfectly well. I entreat you, be sure that the people who govern you also understand that the research, development, and growth that you can accomplish in education is cheap compared to the amount of money that is being spent, or to the amount of damage you are making by not having it, or by having it done poorly.

Professor João de Sales Pupo (Brazil) pointed out a number of difficulties that have been met in Brazil in attempts to teach physics students to think instead of to memorize.

First, he said, as the teachers of other subjects do not accept this view, students frequently oppose it because they are already trained in the old system and are afraid of being unsuccessful in the new one.

Second, the students prefer to be prepared for the kind of formal questions and problems they meet at the entrance examinations at universities in Brazil.

Third, high-school teachers have no time to prepare themselves for their classes (they frequently give twelve classes a day) and have no equipment for experimental work.

Professor Cernuschi (Argentina) pointed out that the use of films in underdeveloped countries may become very important for the necessary massive preparation of teachers. A good film that shows a lecture by a competent physics teacher may have an impact on children in small towns all over the country, raising new interest and making them ask questions of their teacher, thus forcing him to improve his teaching.

THE PLACE OF ATOMIC PHYSICS IN GENERAL EDUCATION*

Mr. John Lewis (U.K.)

Throughout this Conference there has been great unanimity that it is not factual knowledge that is needed from our physics teaching, but understanding. All this has been very general; it is much harder in practice to see how this should be done. In this short paper I hope to

---

*Material presented by the author has been subject to extensive editorial revision.

be down to earth, suggesting practical ways in which our aims can be realized. I hope to make a plea that we could find some of the answers by including atomic physics in our courses.

Far too much school work — and I am considering only school work and not university work — has centered around physics prior to 1895. The American PSSC scheme includes much atomic physics, and we should all be very wise to take careful note of that scheme. Professor Zacharias goaded the United Kingdom for being so decadent as not to introduce the PSSC scheme. He also mentioned with regret that the PSSC textbook had not been translated into French, German, and Dutch. I was not clear whether he included England in this group and whether he was anxious that we should translate it into English!

However, I do welcome this opportunity for explaining why in fact we have not adopted the PSSC scheme, even though we admire so much that is in it. We have a very well established tradition of physics teaching that provides courses extending over many years. We do not require a one-year (or even a year-and-a-half) course. We want a course suitable for the five years from eleven years old to sixteen, a course that could be extended to eighteen-year olds. A one-year course just does not fit. I cannot speak for France, Germany, and Holland, but I suspect that the same applies to them. They too do not require a one-year course that does not fit into their tradition.

There are also other considerations. The intellectual demands of PSSC are such that the course is suitable for the seventeen- and eighteen-year-old children for which it was designed and is not therefore suitable for the twelve- and thirteen-year-old children for whom we require a course. There is another factor: in England a little over 95 per cent of the teachers who are teaching physics in our grammar schools have had a university training in physics. This means that we do indeed have physicists in our schools teaching physics. I am not sure whether this is the case in the United States. I gather the situation there is far less satisfactory.

However, all is not well with our English physics teaching. Our problem is very different. What we need is a change in methods. There has been far too much emphasis on facts and far too little teaching for real understanding. That is why we have now set up in England, under the aegis of the Nuffield Foundation, a programme for producing a scheme similar to the PSSC scheme but much more suitable for English requirements.

We shall provide a text for the pupils, a large teacher's guide, and there will be a laboratory manual. There will be apparatus, which will be of two kinds. We need simple apparatus that the children can use themselves for individual experiments, in fact we believe some of it may even be better than PSSC apparatus! Unlike the PSSC scheme, we shall, however, include quite a large amount of demonstration apparatus. We should very much prefer the pupils to see the experiments themselves rather than to see them on film. Finally, the Nuffield Physics Project is developing an elaborate film programme of its own.

The Nuffield course will also include Atomic Physics as the PSSC course has done. In fact all the new syllabus work throughout the United Kingdom in the last five years has accepted that there should be a place for it. I should, however, like to make a plea that atomic physics has something special to contribute, something that cannot so easily be given by classical physics.

First, in the second half of the twentieth century, when radioactivity and x rays are so much part of our lives, we are failing in our duty to the children if we do not teach them something of these subjects.

Second, the subject is one in which the children are interested and about which they are only too anxious to learn. Professor Boutry has spoken about the difficulty of keeping the children's attention to the simple pendulum. Of course, a good teacher may be able to make elasticity exciting, but it is not normally easy to keep the children interested. With atomic physics there is no such problem. We have developed a very inexpensive cloud chamber that costs less than a dollar and is therefore such that schools can have it in quantity, with one chamber for every three children. One has only to look at the expression on the children's faces as they gather around their own cloud chamber, to see how interested they are.

Third, the subject will necessarily bring in history and the human element. With such names as Roentgen, Becquerel, Curie, Thomson, and Rutherford, how can it fail to do so? Furthermore, the subject provides excellent examples of scientific method, which it is important to communicate to the children.

Atomic physics also teaches children to think and to think critically. In these days of commercial advertising when so much is claimed as being scientific, we have a duty to teach our children to think critically and to sift the evidence. This can come, for example, from seeking the evidence for the particulate nature of electricity, or considering the evidence for the continuous or discontinuous nature of matter. The subject will also raise philosophical problems, as, for example, when considering waves and particles.

In discussions with many of you throughout this Conference I have learned how many are still teaching general science. This usually amounts to skimming the surface of physics, chemistry, and biology with a great danger of superficiality. With atomic physics it is possible to study in some depth without this danger of superficiality.

We hear much about the importance of a link between physics and chemistry. I feel that we, as physicists, have a considerable responsibility to lay certain foundations upon which the chemists can build. A course on atomic physics should certainly include work on energy levels, which in turn will provide valuable evidence for atomic structure that the chemists can use.

There is another reason, a more subtle one, why atomic physics might be introduced. It has not been taught before in most of our schools, and this means we have a chance to suggest the way in which it should be taught and thereby get our ideas across to the teachers. It is much more difficult to get a teacher who has been teaching electricity for many years to change his methods; it is not nearly as difficult to ask him to teach a new topic.

Finally, I should suggest that the teaching of atomic physics in schools provides a fine opportunity for some experimental work of an unusual kind, experimental work that will do much to inculcate the new ideas for which we are all so anxious. Perhaps I might be allowed to give a few examples.

First may I mention an experiment I value highly: the measurement of background radiation. A very simple detector is all that is necessary, something which will give a flash of a neon tube or the click of a loudspeaker, something which can be made for very low cost. I like to put such a detector in front of a class and tell them we are going to learn to count. At first the children will laugh at such a suggestion, but when they count the background for, say, a minute, they will find the human errors involved. Once this has been discussed, they will then discover the variation of the background radiation from one minute to the next. It will make them think about the nature of the readings they are making. They will learn something of the statistics of their observations. By contrast, they can use a simple uranium oxide source with a scaler and notice the statistical fluctuations in that case. All this work will give them an appreciation of scientific readings.

It is important that children should appreciate some of the uncertainties in science. In the past, children have too often been asked to find the focal length of a lens by six different methods or to find the specific heat of brass, of copper, of aluminum, and so on. The children carry out detailed instructions and then ask if their answer is "correct." We need experiments to get away from this.

I like experiments on half-life. Much can be learned from measuring the half-life of thoron, which we can now do in England with relatively inexpensive apparatus. Analogue experiments also have great value here. Line up every child with a coin, let them toss it, and those with heads uppermost sit down. Repeat the process, and they soon get a feeling for half-life. It will also raise the question of what happens to the last child. And so we get them to think. In place of coins, dice are useful as the half-life is just over three throws. We also have a useful apparatus consisting of an inclined board with a horizontal line of holes in it, a large number of ball bearings are poured down it, most go to the bottom and are collected, some fall through the holes and represent those that have "decayed." The half-life can thus be found. The question of whether we need to count the balls will arise. Can we weigh them instead? Do we need an "accurate" balance? Once more we learn about the readings we are taking.

One can go on for a long time with such examples. May I quote a final one. A valuable experiment is to investigate the inverse-square law for a radioactive source. Its value lies in the fact that the law holds for a pure gamma source but does not hold for a pure beta source because of the absorption of the beta particles in air. One can set up these experiments around the room with a source in front of a Geiger tube, half will have a beta source and half a gamma source. We ask the pupils to investigate the law. Of course, at first no one will get a straight line because they will plot the intensity I against $1/d^2$. They will forget the uncertainty about where to measure d, that one has to take into account the thickness of the end window of the Geiger tube. One must not tell them about this at first but encourage them to realize for themselves that they should have plotted $I^{-1/2}$ against d, as this will take into account any zero error involved. The great value of this

is that we are asking our pupils to make an investigation without a predetermined result. How often in the past we have asked them to investigate the validity of Boyle's law? The pupils know perfectly well that
we expect a straight line and will doubtless do their best to give us
one! The value of the preceding experiment is that it will teach them
something of the discipline of a scientific investigation.

In order to teach this modern physics there have been many problems in England about the provision of apparatus. Contrary however to
what I gather was the position in the United States, we have met with
the closest collaboration from our scientific instrument manufacturers,
who have gone a very long way toward providing us with the apparatus
we needed at a price we could afford. Our conclusions are contained in
an interim report that we have now published, and of which copies are
now available.[1] In this report we have also recommended demonstration apparatus as well as apparatus for use by the pupils themselves.
In general, we believe that, as far as time allows, it is better for the
pupils to do their own experiments, but that where such experiments
cannot be done they should be done as demonstrations by the teacher.
Only when this is not possible should we resort to films. To me, it
seems tragic to put so beautiful a piece of apparatus as the Leybold
fine-beam tube on to a film, as the PSSC people have done. It is such
a remarkable experiment, it is such fun to bend that beam oneself, that
we should certainly recommend this as a demonstration experiment. In
England we are fortunate in having teachers capable of handling it.

Lastly, I should like to refer to our film programme. We are producing a whole series of 8mm films in cassette form. It was interesting
to hear Professor Holton's appeal for an awareness of historical apparatus, as two of the films we have already produced — Aston's Mass
Spectograph and Thomson's Positive Ray Parabola — both start with
shots of the original apparatus and conclude with results actually taken
from the original papers. We shall also be producing some films of the
PSSC type for experiments that cannot be done in the classroom.

There is, however, a final type of film that we are now producing
which I do not think has been produced previously. In England we have
a system of categorizing our films: there are those which are suitable
for children, those which children may only see in the company of an
adult, and finally X-films which no child may see at all. These films
are usually horror films or sex ones. We have been making X-films!
I hasten to add that we do not want the children to see them merely
because they are intended for the teacher and go at such a pace that
they would be most unsuitable for the children themselves. These are
films for science teachers, films to suggest methods of teaching, films
on the use of apparatus. They have been made in collaboration with the
Nuffield Foundation Physics Project and have been generously sponsored by the Esso Petroleum Company in England. So far, three have
been made, and others are in production. The first was on the use of

---

[1]Copies of the Report on the Teaching of Modern Physics are obtainable
from Mr. Lewis at Malvern College, Worcestershire, England, or from
the Science Masters Association, 52 Bateman St., Cambridge, England.

centimeter waves in the teaching of optics, the second on experiments in electrostatics, the third an introduction to radioactivity, and we have another almost completed on the use of the Nuffield electromagnetic kit. I should emphasize that these films are merely to help the teacher, and we were merely concerned with showing the teacher how to use apparatus that has now become available in England. I hope this paper has given some indication of the way we are trying to achieve the aims on which we are so generally agreed. I hope that in the months ahead the Nuffield Project will be able to produce something that will at least interest you all.

At the close of Mr. Lewis' paper, he showed the film Introduction to Radioactivity. The Nuffield Foundation Physics Project films for science teachers are obtainable from the Esso Petroleum Company, Esso House, Victoria Street, London, S.W.1, England.

UNESCO AND SCIENCE TEACHING*

Albert Baez (UNESCO)

I want to say something about the role of UNESCO in the teaching of science. UNESCO has to serve more than one hundred member states. It has three large departments: Education Department, Natural Science Department, and Cultural Department. I want to discuss only the budget of the Department of Natural Sciences and to indicate how it is growing. We have the Regular Program, the Technical Assistance Program, and the Special Fund Program. In the Regular budget for 1963-1964 there were 4.9 million dollars. In 1965-1966 we shall have 7.9 million dollars, an increase of 58 per cent. For Technical Assistance the present budget of 4.5 million dollars will be unchanged for 1965-1966. For the Special Fund, 14.9 millions for 1963-1964 is to become 22.9 millions for 1965-1966, an increase of 53 per cent.

These figures are for the entire Science Department, not only the Division of Science Teaching. Considering now the budget for the Regular Program in 1965-1966, 7.9 million dollars, and compare that with the total spent in the last few years in the United States just for one program, the PSSC. The PSSC program has, I believe, cost 8 or 10 million dollars in total, all for the improvement of just one science in just one country. The UNESCO budget for its Regular Program is about

---

*Material presented by the author has been subject to extensive editorial revision.

the same: 8 million dollars. Although this is for only one year, it has to support not only work on the improvement of the teaching of physics but the entire program concerning all branches of sciences of one hundred member states.

UNESCO could never carry out a task of the same magnitude as that that has been done in the United States by the PSSC. UNESCO can never become a fund-granting agency.

What then can UNESCO, with these budgetary limitations, do? It can do two things that cannot be measured easily in economic terms. UNESCO can become a catalyst of activities, and it can serve to put some activities on an international basis. There are certain activities that, if undertaken directly by a single member state, would never have the same international impact as they would if they were done through an international organization like UNESCO.

In the regular activities of UNESCO a certain amount of money is earmarked for the Division of Science Teaching; these funds have been used to support conferences, to bring scientists together, where teachers and others can meet to discuss problems connected with teaching. The other well-known system is that of technical assistance. The program of technical assistance is divided into three parts: one grants fellowships, another provides the service of experts, and the third provides equipment. Of course, with such scanty funds it has been very hard to carry out effective work. Another limitation of our activities is that the technical assistance program depends upon what member states request, and the states that need most assistance are the least able to know what they do need. This has been a very serious problem in the past.

What are our plans for the future? Rather than laying stress on conferences (although well-planned conferences have very great value), other types of activity could be adopted.

Some information on the way it is proposed to develop the science teaching work of UNESCO was given in a paper to the preceding conference by one of my colleagues, Dr. P. Bergvall[1]:

"UNESCO and IBECC are launching a pilot project on the teaching of physics, utilizing new methods and techniques, to start immediately after these conferences and to go on until the end of July 1964. The location will be São Paulo, Brazil.

"The aim of the project is to explore how to utilize films, programmed self-instruction, and simple inexpensive experimentation in high-school physics teaching. At the end of the project, television will be used, and its possibilities in the training of teachers will be investigated.

"We chose IBECC as a nucleus for this activity because for ten years IBECC has been instrumental in the creation, production, and distribution of inexpensive laboratory equipment for the high schools of Brazil. We shall have access to the IBECC workshops, and we have

---

[1]Extracted from a paper "Production of Auxiliary Tools" by P. Bergvall (UNESCO) presented to the First Inter-American Conference on Physics Education, Rio de Janeiro, June 24-29, 1963.

money for salaries for technicians to work there. We also have access to IBECC laboratory facilities.

"We therefore have a unique opportunity for a teacher-training project stressing inexpensive equipment, the use of film, both in the classroom and in the film loop projectors, television, and experimentation on programmed self-instruction to produce manuals and texts.

"The topic will be limited to one part of physics, the properties of light, with a view to discussing the wave-particle duality. This is one of the main themes of the PSSC course, and it will give opportunities of treating classical optics as well as the modern quantum and wave-mechanical approach.

"How we are going to work this project in detail is not yet decided. Speaking in the jargon of programmed instruction, we have several target populations. The most important one will consist of some thirty to forty practicing teachers and professors of physics in the Pedagogical Institutes and universities of Latin America. They will work on the project for the whole year, or at least for the latter half of it. They will take active part in all the aspects of the project: construction of equipment, making of films, writing of texts and programmed instruction manuals. At the end of the course each participant will be bringing to his home country an 8mm continuous-loop projector with films, a kit of inexpensive equipment for experiments and demonstrations, texts, programmed instruction manuals, and a working knowledge of, and we hope enthusiasm for, using these aids in teaching.

"During the last six weeks of the project the participants will be acting as teachers of teachers in high schools of Brazil. It is planned that three half-hour programs will be given in each of the six weeks over a Brazilian television network. The programs will show a "master" teacher on the screen, experimenting, using single-concept films, and so on. We are hoping for a high multiplication factor in this way, and we hope that the participants in the project will be able to repeat the television programs in their home countries."

PHYSICS TEACHING IN CZECHOSLOVAKIA*

Professor M. Valouch (Czechoslovakia)

Since the last world war there have been large changes in the educational system of Czechoslovakia; these have been connected with the building of a new social system in the country. One of the main changes has been to give to mathematics and physics a central role in general

---

*Material presented by the author has been subject to extensive editorial revision.

education. Thus, instead of having only 10 or 15 per cent of the children in secondary school studying science, as used to be the case, all children are now required to have a basic training in science.

This emphasis on the teaching of physics and the need for more physicists in the country have had repercussions at the universities where physics teachers are trained. National education plans fix for each year the number of new students to be accepted by each university in the various courses. For physics this would be based upon the number of physicists considered to be necessary in industry, in research, in teaching, and so forth. At present the medical schools have five times as many applications as the number of available places. In technical schools about one out of two applicants is selected. However, all those who apply to study physics are accepted, as there are fewer applicants than available places.

So far our efforts have been mainly concentrated on widening the educational system. This has put quite a burden in those teaching physics. It has been realized that the lack of applications to study physics at the universities is mainly because secondary schools do not prepare students well in physics and do not develop interest in physics and mathematics. This has led to experiments directed toward the improvement of science teaching. Frequent discussions among science teachers, psychologists, pedagogues, and even parents of students have led already to some improvement. A major experiment has been prepared to start in the next school year in three secondary schools, with the participation of the Mathematical and Physical Society. This will continue for five years, after which period a general reform of the curriculum and syllabus will be prepared.

6

Apparatus for
Teaching Physics

PRINCIPLES OF CLASSROOM DEMONSTRATIONS

AND LABORATORY WORK*

Professor Erik Ingelstam and
Mr. Karl Gustav Friskopp (Sweden)

It is a long time since it was considered possible to teach physics merely theoretically and with blackboard or pencil sketches of experiments. "Away with chalk physics" was the title of one of the first manuals for demonstration apparatus in this century, and since the Second World War the demand for experimental teaching in physics has increased from day to day.

We want to bring out one major principle in our presentation here, and this is simplicity. We do not deny the need for elaborate demonstrations and experiments in higher education, and certainly many of us have begged the authorities for elaborate material for our university students. But let us here begin at the other end, and emphasize the present trend in our schools in Sweden, and certainly in many other countries too, to create simple material of good quality for physical experiments.

The demand for quality is important for several reasons. If the students find the material poor, they will lose interest. Nowadays students demand that educational material, as well as everything else, shall be well designed and constructed in such a way that it emphasizes its function: it must be stable and sturdy enough to be handled by strong schoolboys and hurried teachers. When new material is to be purchased and one has to choose between simpler and more expensive material, one must always think of the cost per pupil who is to use it, or per year, instead of the absolute amount. If a very simple apparatus bought

---

*Material presented by the authors has been subject to extensive editorial revision.

for, say, $20 will last only two years, but you can buy very durable material lasting for twenty years and it costs $100, you have made a real gain.

The best way to teach physics is to let the pupils make as many experiments as possible, and let them work in small groups when they do this. Most experiments should be made by a group of only two students. It is not enough for the material merely to be simple and of good quality if many different experiments are to be made. It must also be inexpensive. One way to get this good material at a reasonable price is to construct one or more sets of combinable components. In Sweden we can get separate sets in various fields of physics, such as basic measurements, mechanics, vacuum experiments, heat, magnetism and electricity, optics, and electronics. We show here one of these sets, mechanics. It was stressed at the Paris Conference by Dr. Roderick that in countries with a less technological background it is particularly important to introduce the simple, basic mechanical elements. The set shown in our exhibit consists of more than a hundred parts, and about fifty experiments can be set up by the pupils themselves from this set, ranging from blocks, pendulums, Archimedes' principle, and volume measurements to stability of equilibrium. Most of the experimental constructions can be mounted on a base plate, so that the construction will be stable. The contents of the set and the manual that accompanies each set will greatly facilitate the teacher's task when introducing basic concepts of physics. Another set is for electronics. With the elements of this set we can perform more than fifty experiments. We have also here a base plate on which we can place the various components. All components are mounted on transparent plastic so that you can see the components themselves, and besides you can also see the symbol of each component.

You might very well object to this last demonstration and say that electronics is far too complicated for ordinary secondary schools. This is true, except for some gifted students who perform some independent work, so that this set is mainly for technical secondary schools.

Our opinion is that these sets, conveniently designed and made, are of great use to the teachers at all levels. A college professor can let his students make more experiments, and with less time-consuming work on trivial details, if the laboratory material is well thought out. But the main emphasis is on teachers at the lower levels of the school system. If such a teacher has access to sets of experimental apparatus provided with manuals for their proper use, he can compensate for much of his own lack of training when he was young. He also saves much of his own time because he is relieved of the task of designing and constructing the material himself.

In this connection, a remark is in order concerning the practical storing of the equipment. In our opinion, the same kind of instruments are best stored in boxes or cases with places fitted for the individual pieces. These units can be kept in wall-cupboards or in some other place easily accessible and controlled. The main advantage of this special method of storage is that it is easy for the teacher to take the cases to the laboratory benches where the students can pick out the elements themselves. More important still is that after work is finished it is easy to check whether the set is complete, and missing pieces can be readily detected. This simplifies greatly the supervision of a laboratory.

Sometimes it is not possible to let the pupils make the experiment. Time may be too short or the material too expensive, or perhaps the teacher wants to draw some conclusion from an earlier experiment. Then classroom demonstration is the natural solution. Of course, the school classroom may vary greatly in size. In Sweden we now have about thirty pupils in each class, and the classrooms are 60 to 70 square meters. If we have much larger classrooms or audiences, other problems arise, which we shall deal with later. The first requirement on a classroom demonstration apparatus is that it is large enough to be seen clearly by all the pupils. In this respect much remains to be done. A meter should have lines of scale division at least 3mm thick, and the whole instrument must be of a corresponding size. An alternative that must be chosen in very large lecture rooms is optical projection. Special mention must be made of shadow projection, which is easy and gives a good result, where a car lamp of about 50 watts gives, thanks to its concentrated filament, images that are sharp enough.

Besides size, illumination is another important factor for good visibility. Great attention must be paid to good illumination. It always pays to equip classrooms with spot-light lamps to illuminate the teacher's desk during experiments without dazzling the students. They can conveniently be mounted in the upper-rear corners of the room, or, if this has not been done, closer to the desk. If good illumination is supplied in this way, smaller scale of the instruments may be tolerated than with medium illumination.

As mentioned earlier, it is sometimes necessary to conduct the experiments in a way different from classroom demonstrations. We can perform the experiments as large-scale demonstrations for more than one class or we can show them with the help of films and the television. Without denying the potential possibilities of good films and educational television, mainly because things can be shown to the class that would not be possible otherwise (for example, industrial applications and research laboratories), we are of the opinion that films and television can never replace good demonstrations. Whenever possible, the teacher, possibly with the aid of some interested students, should set up experiments before the class, and the school should be equipped accordingly. It would be a serious step backwards to rely only on films instead of experiments in any curriculum of physics.

In Sweden we are fortunate in one important respect — we have no serious shortage of good teachers in physics, and so we can use films and school television as a complement. Furthermore, as we have different teachers in different subjects in secondary and higher schools, it is rather difficult to get the television program at the right time. This difficulty could be overcome by films, but there are not very many good films available yet. But as our country has few inhabitants in big areas, we hope to overcome these difficulties and to get good films and television programs. We also hope that films and television will be able to demonstrate new material for the teachers and show them new teaching methods.

Finally, we wish to draw your attention to a possibility that has been used in Sweden in order to extend demonstrations to larger scales and more complex things than can be made in a classroom. In the Technical Museum in Stockholm, thanks mainly to one man, T. Wilner, a series of lectures is arranged, which, on account of their dramatic vigor might even be called "shows." They complement the classroom demonstrations for the classes fortunate enough to go there, for things can be taken up that are beyond the resources of the ordinary teachers in ordinary classrooms with ordinary equipment. The contents range from basic atomic physics, such as is presented in schools, to quantum field theory and relativity.

Most of the experiments and material we brought here have been invented and developed in and for Swedish schools, and we cannot name the persons to whom credit is due. We thank the companies Norstedts School Department, P.O.B., Stockholm 2, Gumperts Skolmateriel, P.O.B., Göteborg 1, for their cooperation, and also Mr. Sten Sture Allebeck, Sveriges Radio, P.O.B., Stockholm 1, Director of the School Television Programs, and Mr. Torsten Wilner, Tekniska Museet, Stockholm No, for their advice and for lending material. Each of these companies and persons has consented to give all information requested and the relevant part of the material presented in this short paper.

In the discussion that followed Professor Ingelstam's talk, one particularly valuable contribution was made by Professor Holton (U.S.A.):

"I want to add one element that has not been mentioned. The word "demonstration" is closely related to "monstrance," some object which is held up during a ritual occasion, such as a lecture. Occasionally it is quite important to bring into the lecture room, if you happen to have an object of this kind, something that really made a difference in physics. We have in our lecture course a few such occasions, perhaps ten or so during the year. When we talk about mass spectographs, we have Professor Bainbridge's original mass spectograph of the early 1930s, and we hold it up. It is a holy object to us; we like it. The students get the feeling that there are objects that are revered because they really worked.

"We have a small beta-ray source that Professor Purcell made for one of his experiments. It is a small brass box, hardly anybody can see it, but I think they can see that I lift it in my hand with some admiration, that it should be such a simple object, and that it did such a good work, and afterwards, before they leave the room they may have a look at it.

"We have a tube of uranium oxide which the Nazi Germans used for one of their atomic reactors and which was found after the Second World War by one of our colleagues; we hold that up, and we pass it around to show how heavy uranium oxide is and that it is not going to kill them to touch it. And one makes a few remarks that under certain circumstances you do not handle radioactive materials. This brings in an element quite apart from what had been discussed so far, which has to do more with the reason why we are interested in physics ourselves."

THE EXHIBITION OF APPARATUS

The organizers of the Conference invited certain organizations to exhibit apparatus which was characteristic of various approaches to the problem of supplying teaching equipment needs. Three main exhibitors were present showing apparatus developed in Brazil, Sweden, and the U.S.A.

The largest exhibit was prepared by the Brazilian Institute of Education, Science and Culture (IBECC) of São Paulo. This organization carries out a program for curriculum reform in science education and produces equipment for supply at cost price to secondary schools and universities. Their exhibit contained equipment for student use, designed to fit the modern idea of simple, inexpensive equipment that leaves to the students the essential part of imagining and doing the experiment, guided with a minimum amount of direction. IBECC has incorporated much of the American Physical Science Study Committee equipment, and all equipment of this type produced in Brazil was presented. Scientific kits produced for student use which could be incorporated into the teaching of science independently of the school syllabus were also shown. These kits have been designed and produced for about ten years for use in Brazilian schools. They have proved to be so valuable to the Brazilian students that frequently they themselves have brought pressure on their schools and on their teachers to acquire the kits. Much interest was aroused among the Latin American participants to the Conference. Some of them were willing to produce similar laboratory equipment in their own countries after their return. Exchange of models and tools was arranged for several of the Latin American nations.

Since about 1930 there has been a considerable activity among the Swedish physics teachers in secondary schools to create material for laboratory work and classroom demonstration. Recently there has been organized a center for developing educational equipment at the Chalmers Institute of Technology in Göteborg. The main principle of the Swedish equipment is that simple units can be combined to form various more complicated sets of experimental apparatus. The units shown in this exhibition covered mechanics, heat, and electricity. For student laboratory experiments, individual units of the same kind are stored in the same box. In this way good order is maintained. The material was placed at the disposal of the conference by the manufacturers, Norstedt School Department, P.O.B., Stockholm 2, and Gumperts School Material, P.O.B., Göteborg 1.

An apparatus-drawings project has been conducted in the U.S.A. jointly by the American Association of Physics Teachers and the American Institute of Physics from 1959 to 1962 with the support of the National Science Foundation. It aims at providing detailed information about the construction of over thirty pieces of new and interesting apparatus for teaching physics at the university level. The apparatus is described by means of shop drawings and detailed construction notes, and was originally developed by the physics departments of some of the leading colleges and universities in the U.S.A. The drawings and notes provide sufficient information for other colleges and universities to have the apparatus constructed locally in their own machine shops. Sufficient notes on experimental technique are provided so that student physicists can build the apparatus under the supervision of their teachers, thereby obtaining additional experience in experimental techniques. A wide range of apparatus was included in the exhibit. There were, for example, mass spectrometers, apparatus for exploring the magnetic field of a circular coil, a large electromagnet for lecture room use, nuclear magnetic resonance apparatus, electron paramagnetic apparatus, and many others. Most of these were for student laboratories, but a few pieces of demonstration apparatus were included. The printed materials, portfolios of drawings and notes, bound volumes of drawings and notes, or individual sets of drawings can be obtained from the publisher, The Plenum Press, New York, N.Y. Some of the pieces of apparatus described are available as kits of components that include the machine parts required for assembling the apparatus. These can be obtained from the Ealing Corporation, 23 Leman Street, London E.1, England.

7

The Historical Approach
in the Teaching of
Physics *

G. A. Boutry (France)

Let us begin with a story.

The first edition of the <u>Principia</u> appeared in 1687. An interval of eighty-three years had to pass before Laplace introduced in the analysis of gravitation a function of space coordinates,

$$V = \sum \frac{m}{r}$$

and showed that its chief property is that it satisfies the equation

$$\underline{F} = M \underline{\text{grad}} V$$

where $\underline{F}$ is the force of attraction acting on mass M (1770).

Eleven more years of work passed before the same Laplace was able to prove that his function V also obeyed, at all points of "empty" space, the condition

$$\Delta V = 0$$

and it is part of the story that he proved it the hard way, using spherical coordinates (1781). Another twenty years elapsed before Poisson extended Laplace's equation to its general form

$$\Delta V + 4 \pi \rho = 0$$

and for the first time noted the analogy with electrostatics (1813). The same Poisson, though, waited till 1824 before tackling the theory of magnetic attraction on similar lines.

All this, seen today and by men who are completely ignorant of the piling up of hypotheses of all kinds induced by the numerous experiments in electricity, seems quite strange: we are dealing here with one

---

*Material presented by the author has been subject to extensive editorial revision.

of the glorious periods in the history of physics, during which genius was sprouting everywhere. Moreover, though Cavendish's work lay hidden and Mitchell's experiments were unnoticed, Coulomb had shown, as far back as 1777 to 1785 that the inverse-square law was valid in the case of electrostatic as well as of magnetic forces.

Let us proceed. When the young Ohm began to work (1825), the main results of Poisson had been published. So had been the book of Fourier on heat conduction and heat transfer (1822), and we find Ohm in a very different frame of mind from his seniors, much struck by the analogy between the conduction of heat and the conduction of electricity. Note that this trend of his seems to have been ignored by both his contemporaries and his early successors; only the experimental part of Ohm's work was then recognized; also, there was no attempt whatsoever to identify or even to compare the electrostatic V of Laplace and Poisson and the emf of Ohm, though the expressions "static electricity" and "moving electricity" or "electric current" had been in use since 1795 at least.

By the way, it was not until 1828 that Green used for the first time the name <u>potential</u> for the Newton-Laplace-Poisson function V. The theory of potential was still far from complete; until 1830 its mathematics were in the hands of the French school, which at the time was an abstract, nonfigurative one: analytical relations were written, figures were rarely drawn — so that, when in 1832 Michael Faraday first spoke of lines of force, he did this in a completely original way, purely as an experimental physicist, in an attempt to do away with "action at a distance" and to replace this uncomfortable hypothesis by a new physical mechanism. Faraday did not think of his lines of force as curves sufficiently defined by the condition that they were at all points orthogonal with respect to equipotential surfaces. It was the lonely, mathematical mind of Gauss that provided, seven years later (1839), the first synthesis of the developing potential theory and introduced for the first time — so late! — the concept of flux. Still, in this beautiful but purely static exposition, the consideration and understanding of Ohm's work were absent. Kirchhoff saw where they fitted, in 1849, and one had to wait until 1855 to 1856 for a practically complete synthesis of the theory, the celebrated paper of Maxwell, in which electrostatics, the flow of fluids, the flow of heat, and the flux of electricity were systematically compared. Here began the architectural period, in which great expounders rebuilt potential theory on a palatial scale, soon to point out (and quarrel about) flaws and holes here and there in the grand structure.

Eighty-three years then — not including the "between Newton and Laplace" gap — and the combined or successive efforts of at least eight men of genius were needed to acquire the familiar (and today elementary) basis of potential theory — not to count the lesser findings and the numerous flounderings of many more talented people. These acquisitions did not occur at all in what we would consider today as a logical sequence. I am only giving here a chronological enumeration of facts; the reconstruction of the underlying psychology and trend of ideas is no concern of this paper, at least for the moment.

To proceed: let us have a quick look at one of the facets of the experimental part of our electrical achievements. The first experiments of Galvani were in 1786, and, as you know, they were an awful muddle in which physical and physiological phenomena are intertwined. Yet, by 1792, Volta had conclusively shown that the electricity involved in Galvani's experiments has nothing to do with life and biology, and he had attributed it to the contact of dissimilar metals. By 1802, Erman had shown experimentally, with the aid of the electroscope, that voltaic currents can and do produce all the electrostatic effects. It took only sixteen years, then, to reach this basic identification, experimentally starting from one of the most complicated borderline experiments that ever was made.

After that comes a gap: many tried, none succeeded in making an electrochemical generator giving a really steady, constant electromotive force. Then in 1822 came Seebeck's discovery of thermoelectricity, and it was this new source of current that made Ohm's experiments possible. Then, regarding the theory of Volta's potential differences, and that of thermoelectricity: stammerings and silence for a hundred years — a period during which two scientific and industrial revolutions had time to take place.

Switch now to 1870. All the theoretical and experimental data were at hand (and had been at hand for the best part of thirty years) that made possible the construction of a really efficient machine for the conversion of mechanical into electrical power and vice versa. A crowd of engineers and scientists had been working on the dc generator and motor. Who succeeded? A mechanic by trade, Gramme, who, when he entered an electrical workshop six years before, was unable to read an elementary schoolbook on physics without the aid of a dictionary.

Move along to 1883. The unfortunate Lucien Gaulard, a humble chemical engineer, wanting enlightenment in the progress of science, went to the Conservatoire des Arts et Métiers in Paris, where Marcel Deprez was then professor of physics. Deprez understood the possibility and the importance of transmitting electrical power at a distance, and he had made a demonstration in Munich one year before. Gaulard was fired by his lecture; he seems to have had a mind of the Faraday type: he almost immediately conceived what he called his "secondary generator," our electrical transformer of today. He brought it out in the technical world in 1884 and was to die of the consequences, destitute and unbalanced, four years later in 1888. He was no theoretical physicist, and when a very respected member of the French Academy of Science who had just written a book on electricity (a good one) went to the blackboard to demonstrate by mathematical analysis that Gaulard's invention could not work, the unfortunate fellow could only answer that it did work. The name of Lucien Gaulard remains so obscure that our standard French dictionary (Larousse) does not mention it in its historical section.[1]

---

[1] The name of Gaulard, however, is listed in the Encyclopaedic Edition of the Larousse dictionary in six volumes.

It is the duty of the historian of science not only to delineate the development of experimental science and to describe the progress of theoretical science but also to explain how and why it all happened as it did: it is my hope that the story I have been telling you will help you to understand the magnitude of such an undertaking. One has to unravel and separate the webs of die-hard metaphysical survivals, to show what embarrassments arose from the very few tenable fundamental or cosmological hypotheses that were conceivable at the time considered. One has to recognize that there are several distinct types of scientific mind; between the antithetic and complementary Faraday and Maxwell, de Broglie and Schrödinger, a continuous spectrum of scientific intelligence is spread out for us to examine. Is this enough? By no means: one has to take into account the philosophical and ethical attitude of each of the men who played their part in the development of science. Two poles again are found: one attracts the minds that thirst for an explanation of the world and the other aggregate, those who look for order in the world, regardless of what this order may mean. Thermodynamics was built by this second tribe, atomic theory by the first. Here again, all intermediate attitudes of mind exist in the scientific world and will continue to occur till we men become extinct. Last, but not least, when one has succeeded in taking all this into account, one must be reminded that scientists are men and men of their time, with their character and temperament, their heredity and upbringing, their capacity for love and for strife, their sociological context. All playing a role, small or large in the formation of their scientific attitude of thought. Would Clausius have developed his ideas in the same way if he had not been living in a Germany that was in the process of being flooded by the philosophical and metaphysical doctrines of Fichte and Hegel? With Sadi at his knee, Lazare Carnot, educator of his two sons, revered by them, writes Principes Fondamentaux de l'Equilibre et du Mouvement. In 1824, Sadi Carnot writes his Réflexions sur la Puissance Motrice du Feu, a booklet that is generally considered as the prototype of the isolated, entirely original work. It can be shown, however, by a comparison of the two texts, that there is a large probability that, in Sadi's mind, the Réflexions were an appendix to the father's book, introducing for the first time a new form of energy, but keeping and extending to it Lazare Carnot's dialectical substratum: Carnot's principle may have been a monument of filial love — and I have a suspicion that Volta, in his brilliant destruction of Galvani's conclusion was motivated by a dislike for Galvani the man and Galvani the philosopher. Hate also has some creative value. The dislike of Laplace for Fresnel is another example; it hastened the success of the wave theory of light.

Why go on? One understands that in human hands no unique, true, authoritative history of science is possible. The handicaps of the historical scientist faced by the facts are, when all is said, the same as those of the historian proper. There are, there must always be, several aspects to any evolution or revolution: Histoire des batailles, Histoire des idées, Histoire des peuples, Histoire économique.... We therefore will, in the best of cases, have several histories of science, impossible to synthesize completely.

Even if it were not so, even if the complete puzzle of the advent of electricity and electrotechnics could be assembled, it would still exhibit holes and flaws here and there. Throughout the history of science there exist well-established and recurrent circumstances and facts that nobody has explained satisfactorily. What, for instance, of the apparent stupidity which seems to inhibit from time to time scientists of talent, and even of genius, when applications are converned? Inventors and engineers, scientists and professors, they are two different breeds, and though they know perfectly well that they are complementary and should work together, they have always been reluctant to do so. The first telescope, the first microscope, the first vacuum pump, the first dc rotating generator, the first electrical transformer, photography and color photography, the first superheterodyne have not been created by physicists. They all have been breaks through the (otherwise fertile) conservatism induced by specialized logic, although for one good, inventive lead, hundreds of crazy ideas have been proffered. In the laboratory where I have the honor to be Director of Research, we strive to keep one or two tame inventors living with our orthodox and genteely bred research workers; they have never survived longer than a few years.

A queer food for thought is history of science, such as we are able to make it. Sweet and sour, true and misleading, toxic or fortifying according to who eats it, depending on one's strength and breadth of meditation. How can it help our students?

Roughly speaking there are three stages of physics teaching, and to each of these stages correspond somewhat different methods by the teacher and widely different states of mind induced in the successful pupil. I shall call these three stages initiation, graduation, and mastership; it should be understood that these words do not necessarily have the same meaning as similar expressions used in university curricula.

Here, is a very rough synopsis of what happens during a course of physics teaching. It is oversimplified as tables of this kind are. I do not like it, and every one of you could with some little thought draft a better (and more complicated) one. Such as it is, it will serve my limited purpose.

Initiation

This first process lasts of course several years, during which the relations between pupil and teacher are in a state of constant evolution. However, and whether we like it or not, it is impossible to get away from the fact that during this period the teacher has to "sell" his subject to a totally unprepared group of boys and girls (generally a large one). One might object that today physics is everywhere, and that this group has been hearing for years, through the press, radio, and television, about electronics, nuclear energy, and space travel. This is true, and in my humble opinion, it makes the situation worse: when, in days long gone by, I first entered the physics room in my school, my state of mind was completely virginal; I knew nothing of what physical science was and of what it could achieve. Today, your adolescent

Knowledge Acquisition versus Psychological Evolution During a Complete
Course of Physics

Stage	Curriculum	Teacher's Methods and Intentions	Successful Pupil		Number of Pupils
			Acquisition	Frame of Mind	
INITIATION	Basic facts and laws covering the whole field of physics (frontiers vary) ──── Schematic outline of theories	Slightly dogmatic teaching. Emphasis on quantitative as opposed to qualitative experimentation ──── Absence of any philosophical leaning	Storage of facts and laws ──── Initiation to the art of systematic experimentation	Noncritical ──── Desire to know more (the questioning stage) ──── Attempts at personal experimentation	Large
GRADUATION	Reduced; e.g.: The triad electricity, optics, thermodynamics	General classical theory used as a guide ──── Systematic examination of facts and measurements (errors, importance of second-order terms) ──── Indication of theory failure and introduction of modern ideas	Good frame of systematized knowledge ──── Habits of thought and hands established	Plentitude ──── Admiration for the theoretical architecture ──── Seeds of doubt sprouting	Reduced
MASTERSHIP	Reduced once more: e.g.: description of the gaseous and solid states of matter	Systematic expounding of current theory and comparison both with facts and other theories ──── Attempt at a completely honest picture of present-day knowledge	Versatility of reasoning and thought ──── Initiation to teamwork	Constructive criticism and creative doubt Scientific philosophy as distinct from metaphysics	Small

audience comes to the class prepared to learn at last how the marvels they hear about every day are achieved... and you have to talk to them about ebullition and the swing of a pendulum. It takes talent indeed to hold securely and for a long period the attention of most of such a group.

At this early stage is help forthcoming from the history of science? To a limited extent, certainly. One has to confine oneself to anecdotes, of course, but a few of these should prove helpful. The story (it may be true) of Galileo Galilei daydreaming in the Baptistery of Florence and rhythmically nodding to the swinging of a lamp there; Galileo behaved as an exceptionally bright child would, and you may emphasize that the real difference lies in the fact that he measured what everyone could see.

At a later stage, more elaborate stories can be told, typical of which is the expounding of Lavoisier's work on oxidation: the trend of this is both beautiful and simple. At the end of the initiation period, the tale of the discovery of the laws of electromagnetism might serve a purpose: how only twenty days passed from the arrival of Oerstedt's letter in Paris to the almost final account of the calculations and experiments of Ampère, Laplace, Biot, and Savart; how, year after year, the same Ampère must have produced induction phenomena and remained unaware of their existence till Faraday came upon the scene.

One must be truthful, of course: no pupil at this level could differentiate between an "arranged" story and a downright lie. At the same time, care must be taken not to introduce doubt in the students' mind at too early a stage: first, teach what is firmly established before any systematic skepticism is introduced. Any attempt to proceed otherwise might create in the pupils a negative bias which could become detrimental to their further progress. However, it will not do to paint men of science in the garb of infallible beings: steering between these two rocks is not always easy.

Another pitfall awaits the imprudent user of historical examples; this is an age when our pupils (happily) receive experimental as well as ex-cathedra teaching. The chances are, that the account of a celebrated experiment in front of a bright audience will produce a request to see the experiment in question. Of course, in most cases, one can back out from this situation by explaining what is obsolete in the original setup and by executing a modernized version; if well done, this maneuver will teach the class something additional. There are also cases where the situation does not admit this side-stepping. Then what happens? Let me tell you a story which still rankles though it is now forty years old.

I was then a teacher with six months' experience in a French lycée. The second year in our curriculum was elementary electricity and optics. Electromagnetic induction was the subject; I had successfully introduced the notion of magnetic flux when somebody asked me, at the close of a prepared demonstration, if induction could be demonstrated using the only earth's magnetic field; this, I think, shows that the boys were really following well. I answered yes, of course. "Then one has to move a coil in a certain way, Sir?" This started me on the story of how, in 1845, some physicist, whose name I have now forgotten, demonstrated induction by the earth's field by simply crushing in his hands a

single loop of wire. There was an immediate outcry for the experiment
to be performed. Now, the only galvanometer which we had was a sturdy
but very insensitive wall apparatus, quite unable to show anything but
the meanest kick in an experiment of this kind; with boys of this age,
a gentle tremor of the spot will <u>not</u> do. Now, thanks to the arrangement of the syllabus, the galvanometer was still just a box to these
boys: description, calculation of sensitivity, etc., ...were to come later.
I had three minutes left and rashly decided to plunge. The loop I crushed
in my hands as quickly as I could and ... the spot swung nearly to the
end of the scale. The audience triumphally departed, leaving me to the
sudden remembrance that I had left a large bar magnet in the drawer
of the table, just under the galvanometer leads. What would you have
done in my place, at the next lecture?

Graduation

At the next stage, the teacher bears the name of Professor; he and
his assistants lecture and demonstrate to students who have freely
chosen to include physics in their curriculum. The whole setup and
atmosphere are different. Nobody has to be attracted. All are prepared
for a long and steady pull. The Professor is expected to deliver a solid,
logical, and monolithic exposition, in regular doses and, in fact, strives
to do precisely that. There is nothing else that he can do. The physics
of yesterday, the physics of continuity and certainty, is also the physics
of everyday. It has to be taught and learned if the student is to acquire
a general background without which he could not later specialize and
reach a useful level. Therefore, classical thermodynamics, electricity,
and optics will be developed, the professor steadily building in front of
his students the noble architecture that is the work, the achievement of
the second half of the nineteenth century. In other words, two fundamental questions will not be raised when the course starts. What is
space, what is matter? Of course, the professor will establish safe-
guards, at least in what is contained in the second of these questions;
he will mention atoms and molecules, only to dismiss them as unnecessary encumbrances for the study of macroscopic effects. Later he
will summon them again to his aid, for instance when attacking the study
of electrolytic conduction or of ionization, or in giving an introduction
to the kinetic theory of gases.

It is very possible that not everybody here is prepared to agree that
this is the way which should be followed at the beginning of the graduation stage. I know that many attempts have been made, and are made
today, to start from atoms, molecules, ions, and electrons and recon-
struct classical physics with these materials. I am not sure that the
difference between the two methods is not almost purely formal. After
all, the mathematical formulation used in all macroscopic physics
remains the same in both cases.

Which method is used, however, is not important for our purpose.
What can the history of science bring to the lecturer at this stage of
education? What the professor has to do is to take his student, probably
in two years, along the road that took a century and a half of meditation
and experiment to build. Therefore, he is bound to leave unexplored

many inviting forks of the road and to follow a logical line. He will never attempt to unfold his tale in the historical order. In fact, he cannot do it. Think about the notion of energy, for instance, and of the state to which he would reduce his audience if he were to consider Carnot's principle before introducing the principle of the conservation of energy.

The easiest approach is through men and their psychology. What makes a physicist? What is his mental and sensorial equipment?

Here we have a direct follow-up from the anecdotes used in Initiation; but one can probe deeper and try to describe some of the types of mind met with in the history of science. Contrast and parallel, to take only one instance, the minds of Ampère and of Faraday — the one introspective and analytic, looking always inwards, the other intuitive, thinking in pictures, always looking around him — or those of Sadi Carnot and Clausius or of William Thomson and James Clerk Maxwell .... Slowly, some conclusions will emerge: They will not yet be of high philosophical value (I shall come back to this in a moment), but they will serve: the facts have to be sifted, scraped, and cleaned, and shown to be reproducible in severely specified circumstances before they began to look like the phenomena described in the books and in the lectures; that "accidental" discovery is a myth which must at all costs be exploded (think of the discovery of electromagnetic induction, of Hertz stopping in his study of electromagnetic waves, for an investigation of why the behavior of his spark-gap oscillator changed according to whether the shadow of his colleague thrown by a mercury vapor lamp was projected on or off the oscillator, and thus "accidentally" discovering photoemission; of the man who, in Crookes' laboratory put some photographic material in the drawer of a table on which a cathode-ray tube was installed and, complaining that it was spoiled, was told to go and store it somewhere else; and of Roentgen discovering some time later the same curious trouble but systematically investigating it...); that creation of any kind means strife, and pain, a capacity for concentration, for meditation, and for constructive patience.

But for our bright students this approach will not suffice: the story of ideas must be tackled. Now is your chance to start showing that man is an animal of few ideas and that, in his search for "explanation," he always has been oscillating between a fluid and a continuous universe and a cosmos full only with discrete, minute units. Tell your students if you can, if you dare, the story of the foundation of thermodynamics, with what pains and labor, and how late the concept of energy was finally acquired. Tell them how the correct formulation of the First Principle arose not from the picture of a fluid and continuous universe but, on the contrary, from the hypothesis that this whole universe consisted of atoms connected by central forces and how the combined efforts of Joule, William Thomson, Clausius, and Helmholtz were needed to reach it (1847-1859) though Dr. Robert Mayer, working at the start from physiological phenomena had given a metaphysical statement of conservation of energy some years before. Explain how, in a notebook of Carnot, dating between 1824 and 1832, which nobody had bothered to read, was discovered (in 1878.) a statement of the First Principle and a description of the celebrated experiment that Joule and Thomson were to reinvent and successfully perform in 1854. Then dwell for a moment on the fact that the point of view of Carnot in the Réflexions sur la Puissance Motrice du Feu and the whole of his reasoning are completely abstract in their nature: he was not concerned with any universal hypothesis, not concerned with any explanation. He was seeking for order in the universe. Show how in the second half of the nineteenth century this attitude of mind prevailed and slowly ebbed over the whole of classical physics, to make it the proud construction which you are describing in your lectures. Your best students will begin to understand that the palace is only a front, a "façade," and will be prepared for what must follow.

In the second part of his graduation course, the professor has of course to change his point of view, and it is the historical approach which will enable him to do it. The most convenient turning point is probably the chapter on black-body radiation, the integral laws of which can be deduced from the two principles and from them only, while the analytical law cannot be so deduced. Care must be taken here to show clearly the meaning of this fact: no structural hypothesis whatsoever is needed for the derivation of Kirchhoff's, Stefan's, and Wien's laws; but when one wants to tackle the spectral distribution of black-body radiation, one must introduce a structure of matter hypothesis, that is, describe microoscillators; whether you are led to the erroneous formulation of Wien and of Rayleigh or to the correct law of Planck depends only on the properties that you assign to these oscillators. After this first point is well understood, the second point must be made that Planck's law can be derived only by endowing the microoscillators with properties that are in direct contradiction with the laws appearing to govern exchange of energy in the field of macrophysics. Now is the time to introduce the laws of photoemission and the Einstein equation and to show that the photon hypothesis is necessary to explain both the emission of radiation by matter and the extraction from matter of electric charges by radiation.

Now this utilization of history of science has its difficulties. The main snag has two aspects: there exists no good book on the development of physics in the nineteenth century, and your standard professor of physics (graduate course) is seldom equipped with the historical and philosophical preparation that will enable him to navigate successfully in these waters: he has a few other things to do. It is for these two reasons that it will generally not be possible to go very deeply into the minds of physicists of genius and into the story of the development of ideas. But some attempts must be made: a graduation course would not be the same in its absence.

Mastership

The guiding of the graduate student into mastership would not be the same if some help has not previously been forthcoming from history of science. This is the period of teaching during which the historical approach merges into the philosophical, exactly as contemporary history merges into present-day politics. Great care must be taken and much work done to ensure that the landscape of modern physical theory shall be surveyed from a correct angle and from a sufficient height. In other words, your admiring pupil of yesterday, taken first through the ruins of classical theory, then through the old figurative quantum theory, to enter finally the cold, nonfigurative universe of today, and to knock at the door — closed for the present — of universal field theories must not emerge from the ordeal convinced that it is all a sort of esoteric game in which the wildest surmises are permissible, in which you make your own rules of play, in which you can disregard blatant contradictions and evident failures to cover part of the ground that has to be covered, and so on.... This is precisely what is bound to happen if you do not inject from time to time a little philosophical thinking into the curriculum.

A quick survey of the ideas underlying classical physics and the old quantum mechanics is, at this stage, easy to undertake, and it will at once show that three basic metaphysical ideas underlie these two structures, to wit:

1. The universe is intelligible.
2. All the laws governing phenomena take their place in a single general physical theory.
3. This general physical theory can be deduced from a set of hypotheses that forms what the human mind considers to be an "explanation." This explanation is unique.

One should dwell a little upon the third proposition and show again how, since the origin of natural philosophy, thinkers have always alternated, oscillated between two "explanations" of the world, the first involving continuity and certainty, the second atomistics and probability; that in the history of physical thought, humanity can be depicted as climbing slowly the spiral staircase inside a tower. From time to time, windows allow the view of two landscapes, surveyed from an ever-increasing height, widening, never merging one into another, and also never complete.

As soon as this has been explained and understood, it seems to me that the next step is inevitably indicated. It is to show that the modern waves and quantum mechanics appear as a challenge of the exactitude of proposition 3 in its present form, as a deviation from our helical ascent. We seem to have at long last reached a platform from which we see at once, merged into a single whole, large sections of the alternating landscapes of old. That a price has had to be paid for this achievement now appears normal; that this price consists in ceasing to consider the concepts of continuity and discontinuity as opposed and irreconcilable but rather as complementary will not seem scandalous; that the reason why proposition 2 seems today shaken is that we must grow out of our present too-narrow rules of thought and state of mind. That a new and wider scientific logic and new mathematical formulations must be discovered and established will seem evident. Isn't this what is needed to inspire onwards the young thinkers whom we strive to prepare, to endow them with the necessary constructive criticism, to convince them that the continuing metaphysical struggle between determinism and uncertainty is now an affair of old men?

The history of science which we should have always at our elbow —
the history of physics from the death of Isaac Newton to the formulation
of the Louis de Broglie hypothesis — does not exist today. Would it not
be a proud undertaking of the Commission on Physics Education to fos-
ter its birth, to organize its development? We need to be taught before
we teach others.

Index

www.ingramcontent.com/pod-product-compliance
Lightning Source LLC
Chambersburg PA
CBHW070408200326
41518CB00011B/2109